Santa and His Friends: Carving with Tom Wolfe

Text by Douglas Congdon-Martin

1469 Morstein Road, West Chester, Pennsylvania 19380

Library of Congress Catalog Number: 90-61513.

Printed in the United States of America.
ISBN: 0-88740-277-1

Published by Schiffer Publishing, Ltd.
1469 Morstein Road
West Chester, Pennsylvania 19380
Please write for a free catalog.
This book may be purchased from the publisher.
Please include $2.00 postage.
Try your bookstore first.

Contents

Introduction 5
Carving Santa Claus 7
Painting Santa Claus 29
Carving a Robed Gnome 45
Painting a Robed Gnome 53

Introduction

Santa Claus, elves, gnomes, wizards. These fanciful characters delight and mesmerize us. All of them are magical creatures, possessing great and wondrous powers. Santa is of course the power of good. With his all-seeing eyes the "jolly old elf" espies the good deeds of children. Then, when Christmas rolls around, he makes a whirlwind tour of the planet to reward them with their fondness desires.

Gnomes and wizards are not always so benign in their magic. They are impish at best, and sometimes even sinister. Yet they fascinate us too. Whether they are Snow White's dwarves or King Arthur's Merlin, they have made their way into our stories, fairy tales, and, of course, our imaginations.

Tom Wolfe has a bit of this magic himself. In his hands a piece of wood, carved with the simplest tools, is transformed and takes on a personality of its own. It comes to life.

Tom also has the power to share his magic with others. He has done so in several previous books and he does it again here. The projects he shares in this book are perfect for the beginning carver, or the person who likes to whittle for fun and relaxation. They also give the opportunity for the more experienced carvers to be creative and expand their repertoire. And whether experienced or not, the objects in this book are just plain fun and make wonderful gifts.

Tom's workshop is at Mystery Hill in Blowing Rock, North Carolina. He sells his original work there, as well as reproductions of his most popular carvings. These reproductions are finding fans around the country. Tom's wife, Nancy, is a partner in this venture and they also join their talents to create country dolls.

Tom shows his work at many craft events around the nation, winning recognition for his unusual talent. He has shared this talent in several books published by Schiffer Publishing. Previously published works include *Country Carving*, *Pig Pickin'*, and *Country Dollmaking* (with Nancy). Simultaneously with *Santa Claus and His Friends*, Tom is offering *Carving Bears & Bunnies* and *Country Flatcarving with Tom Wolfe*.

The carvings we are doing in this book can be done with the simplest of materials and tools. A piece of 2 x 2, 3 x 3, or 4 x 4 in white pine, spruce, or clean fir, and measuring anywhere from two to ten inches long, is all that is needed for the body of the Santa, gnome, or wizard figures. The thicker the wood and the shorter the figure the more plump it will appear. The thinner the wood and the longer the figure, the more lean. A good, well-sharpened carving knife is the best tool, but a sharp pocket knife may also be used.

The only real secret to carving these characters is that the wood should be square or nearly square. This is important because the dimension from corner to corner will give the carving its depth and width.

We hope you will have fun with the projects and will use them as a starting point for your own magic and creativity.

Carving Santa Claus

Sharp tools are very important. Take a few moments to hone your tools to their best edge before beginning.

The length of wood you use depends on the desired shape of the character. The shorter the wood, the plumper the figure. The longer the wood, the more lean it will appear. Exact measurements are not important in this carving. Instead the measurement will sort of emerge from the wood as you go.

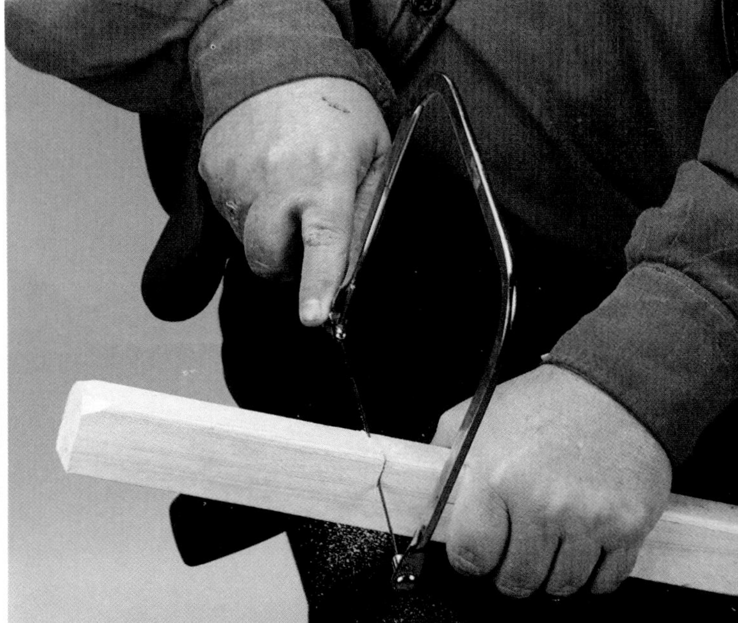

Use a coping saw to cut the piece to length.

The tip of the hat will be positioned to one of the corners.

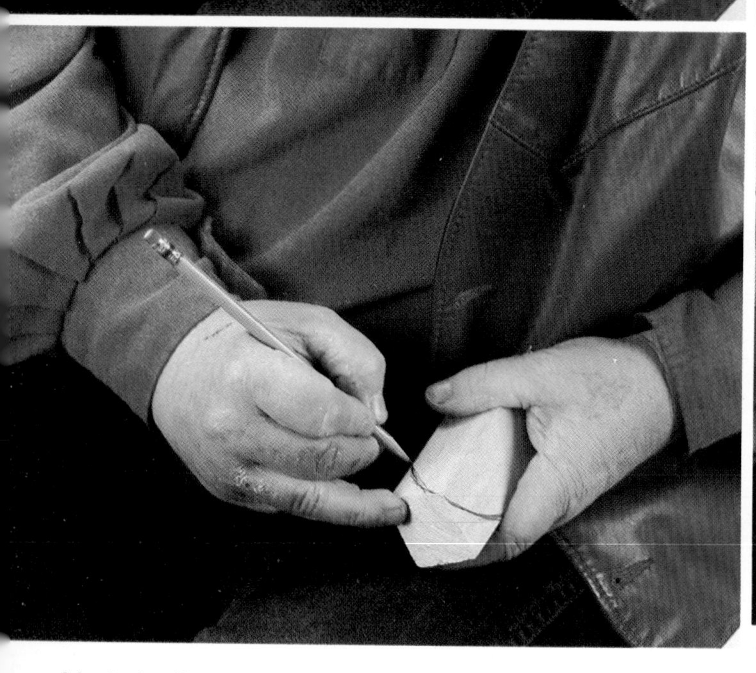

Mark the line of the hat on two sides.

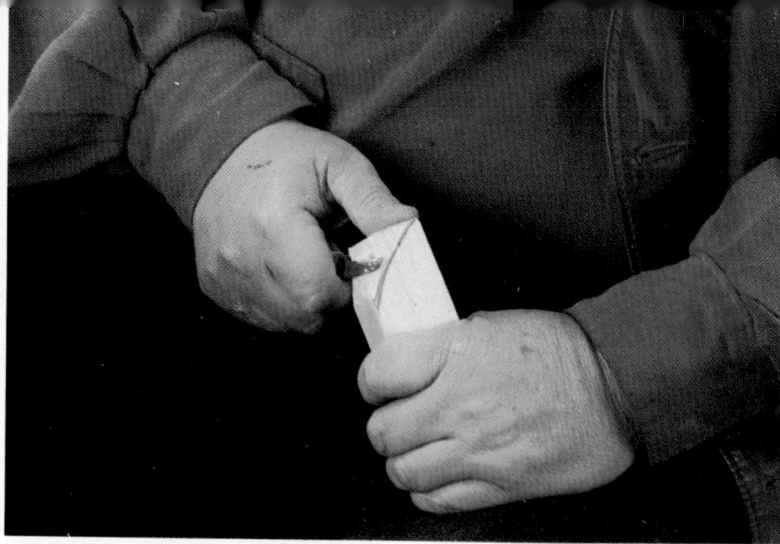

Remove the excess wood.

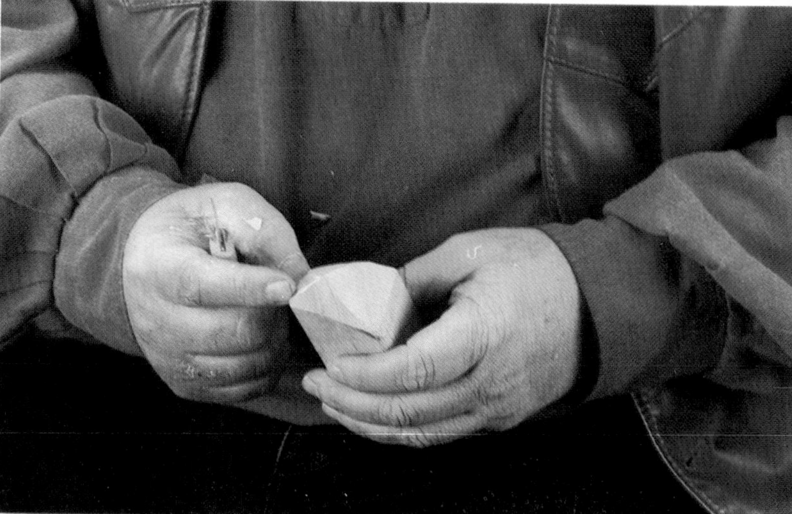

Begin to visualize the point as the tip of the hat. Visualizing helps you end up with the character you want.

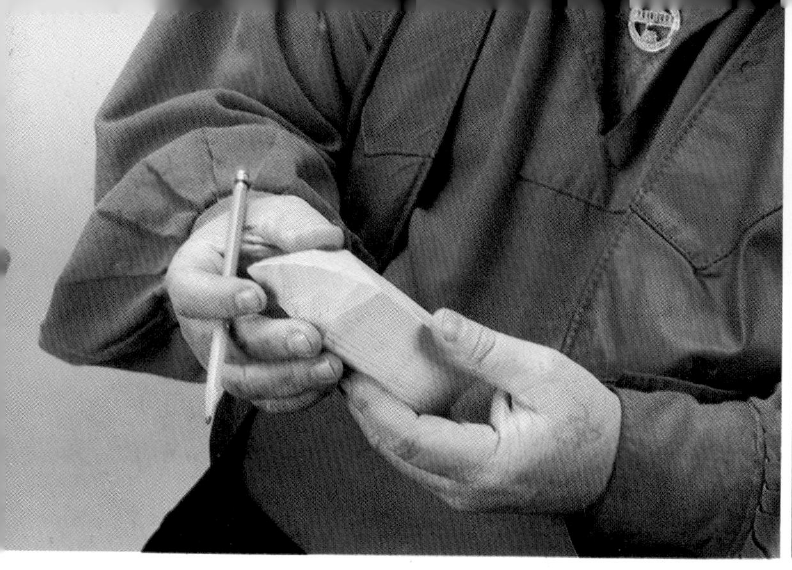

At this point you need to decide whether the character is going to look up or straight out and determine how long the hat should be. In this case it seems Santa should be looking up, perhaps checking to see if the sky is good for flying.

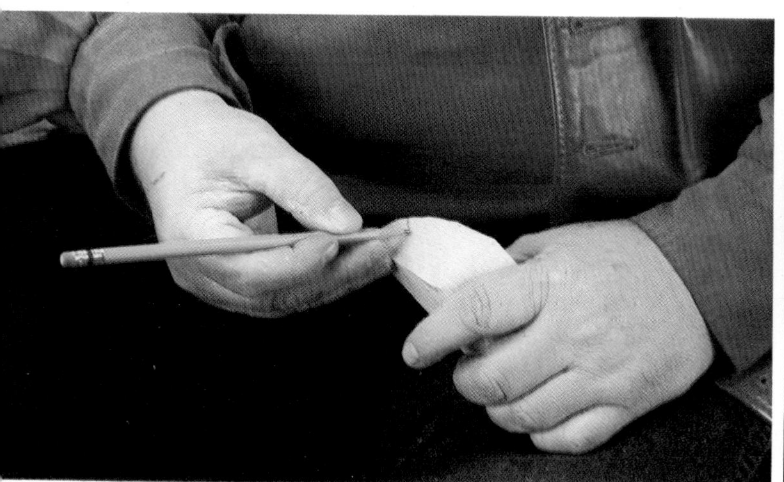

Mark where the hood meets the forehead.

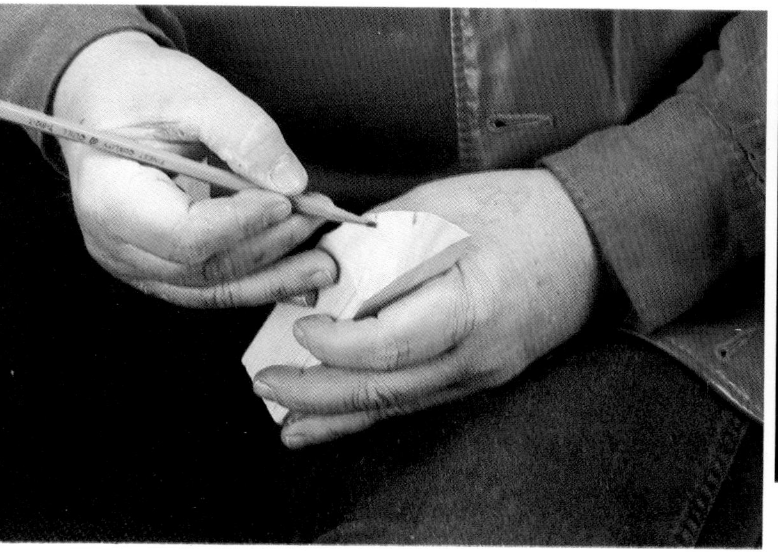

Mark the bottom of the nose...

and the bottom of the beard

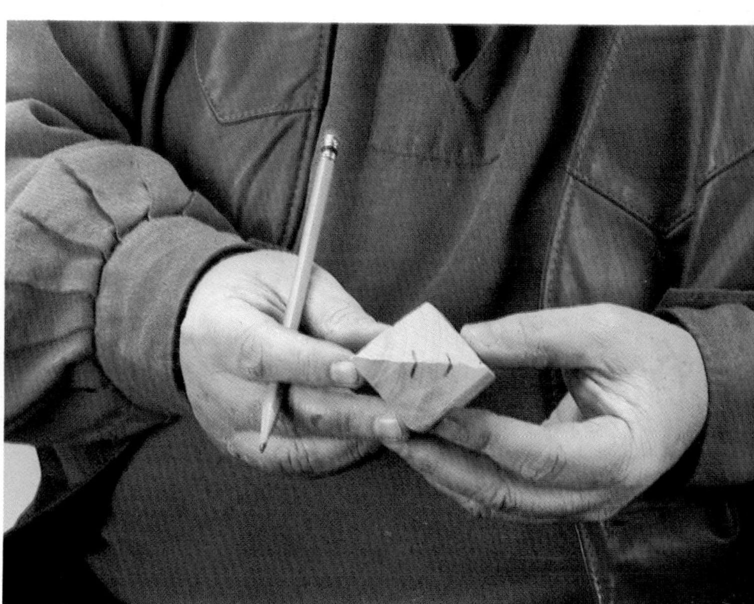

Now we're beginning to see the advantages of the square shape.

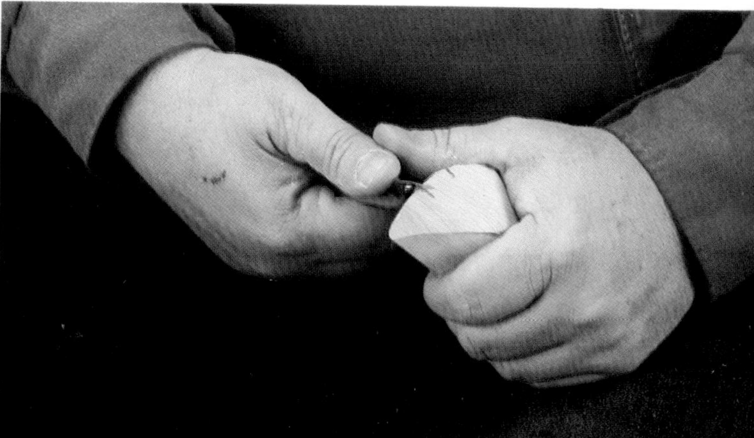

Cut a stop line where the forehead meets the hood. This is done by cutting straight into the wood along the line. By carving back to the stop line, you avoid the danger of overcutting.

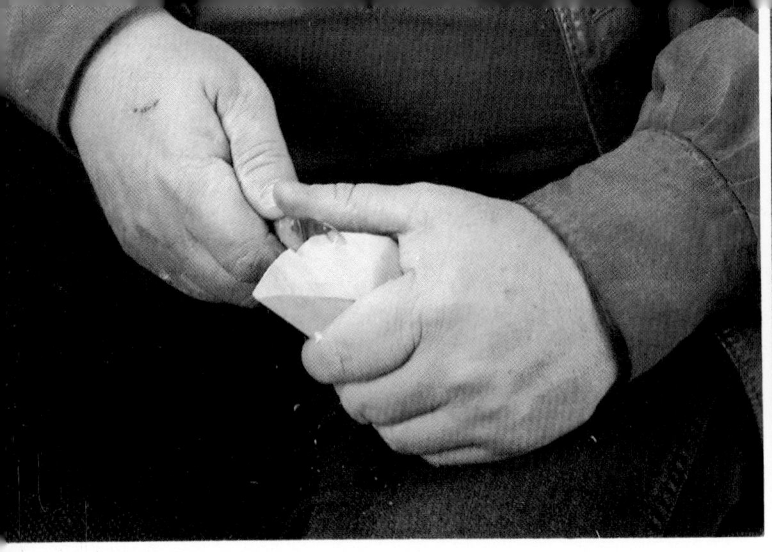

Cut a notch back to the stop and cut another stop at the bottom of the nose. Cut a notch back to that stop.

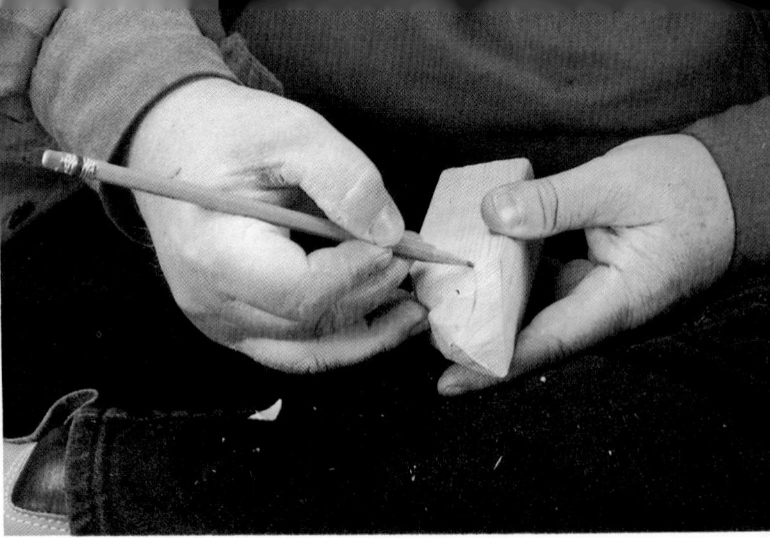

Draw a line marking the edge of the hood...

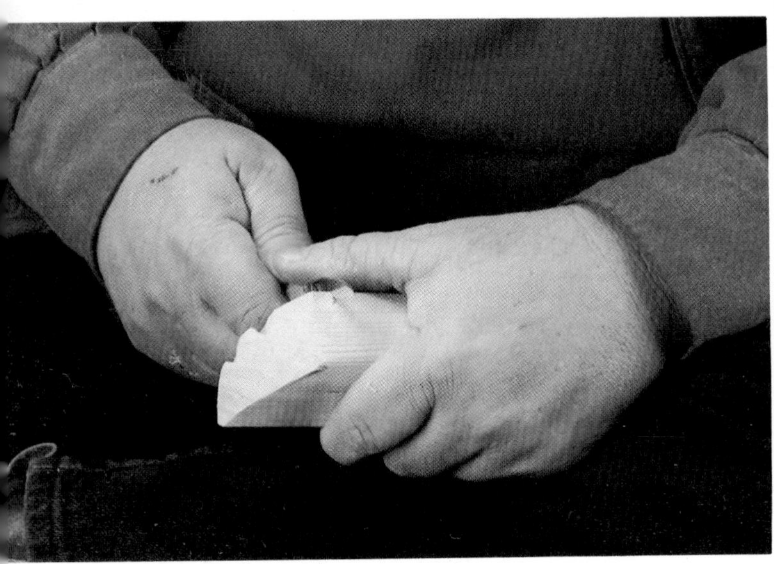

Then cut a stop at the bottom of the whiskers

on both sides.

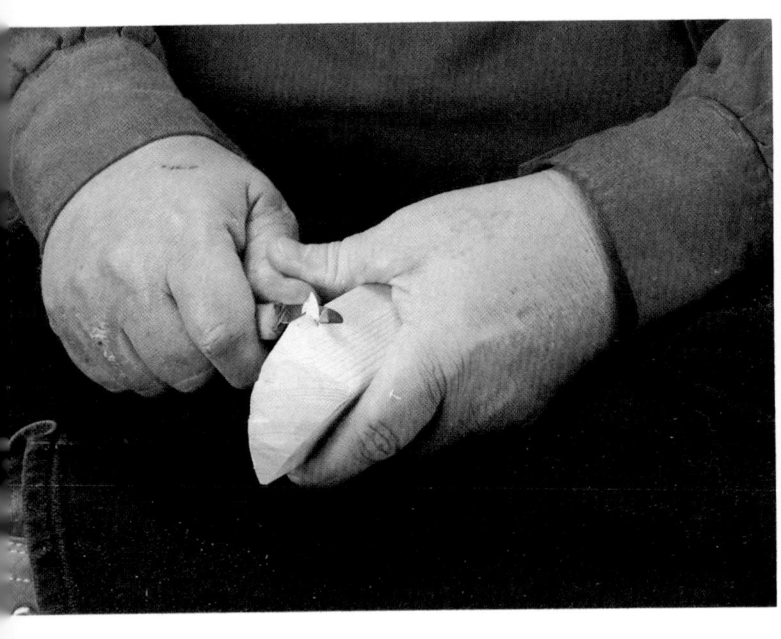

Notch back to it.

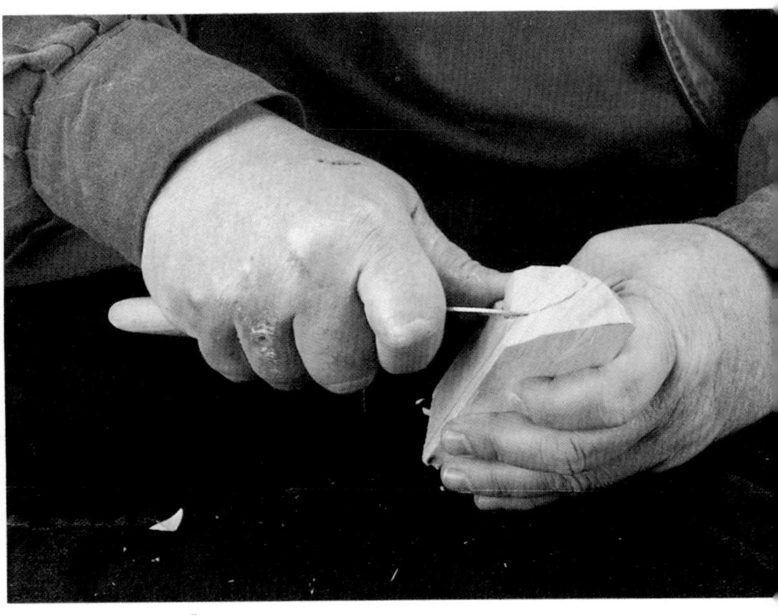

Cut a stop on the line...

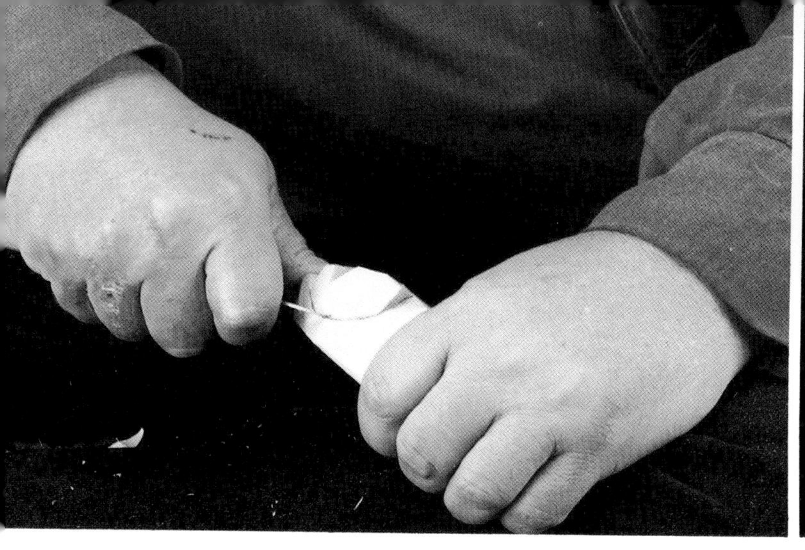

to delineate the face and give it more depth.

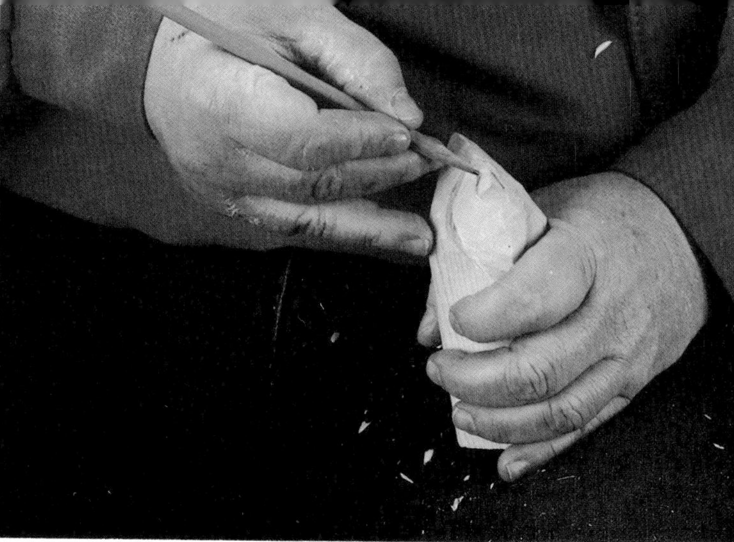

Define the nose by marking it with a pencil.

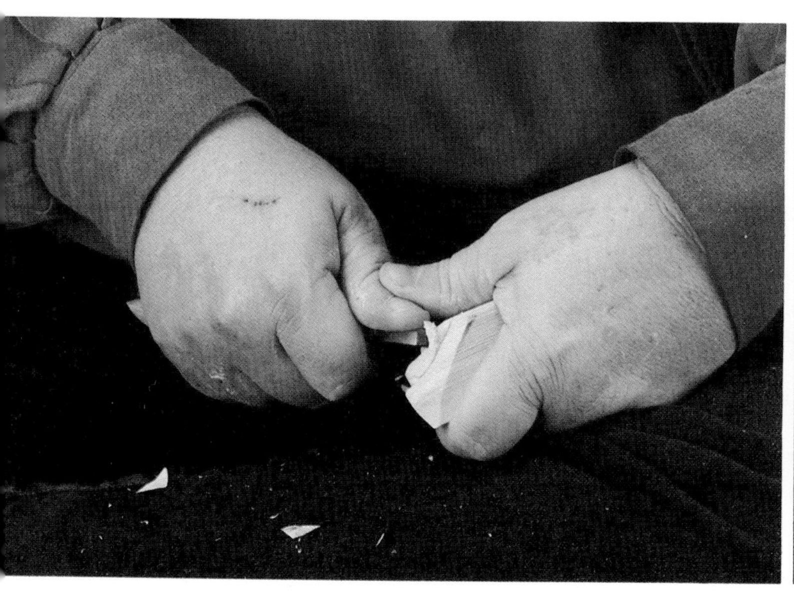

Cut back into the stop to notch it out.

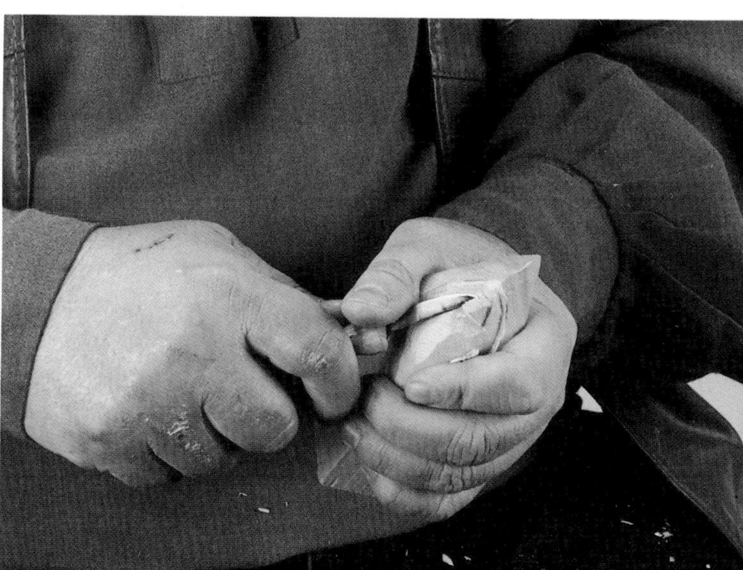

Carve back toward the nose, leaving the sockets for the eyes as you go.

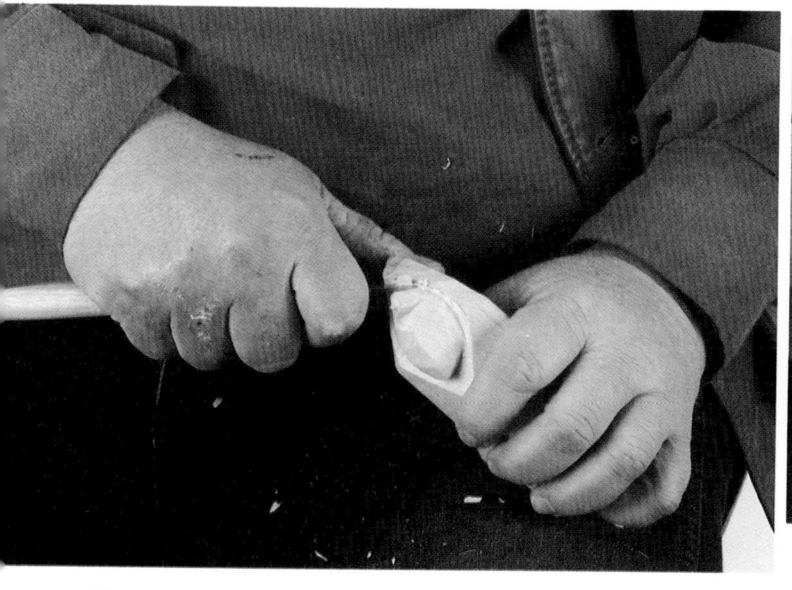

Finish the silhouette of the face all around and refine the lines.

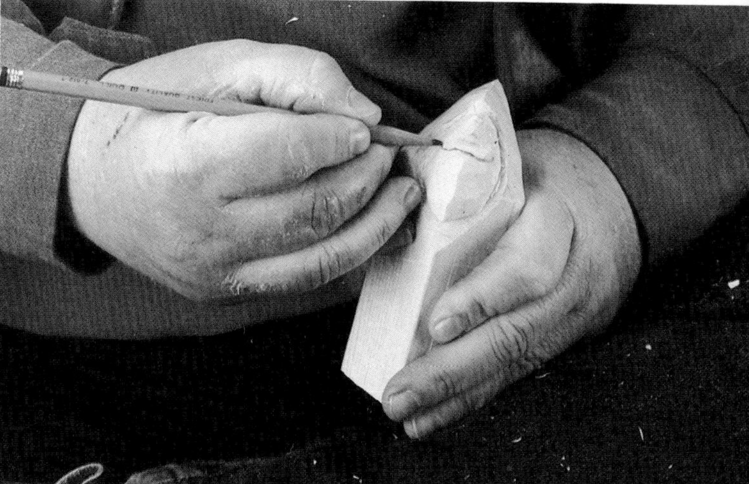

It is time to decide the shape of the moustache and beard. In Santas the moustache should go up in a friendly sort of way. In wizards it might go down, giving a more sinister look. Mark the bottom line of the moustache.

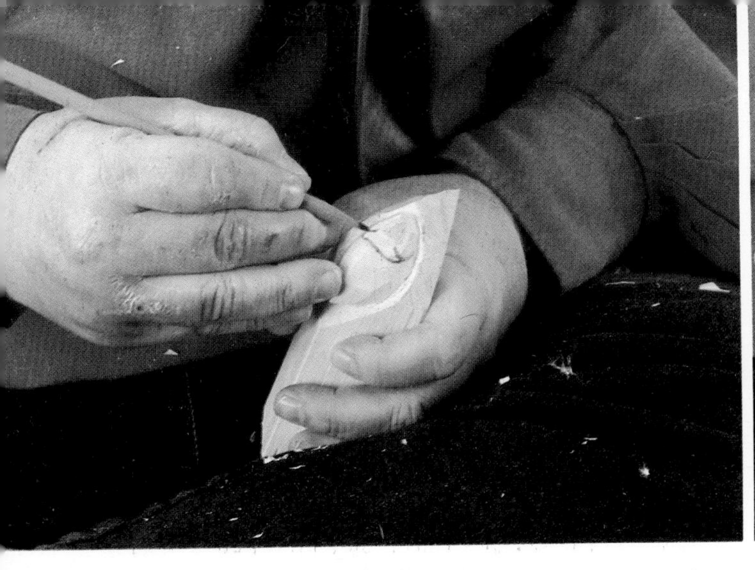

Also draw a line from the nostril to the end of the moustache to define the top edge. It, too, should turn up in the Santa. It is also the line for the bottom of the cheek.

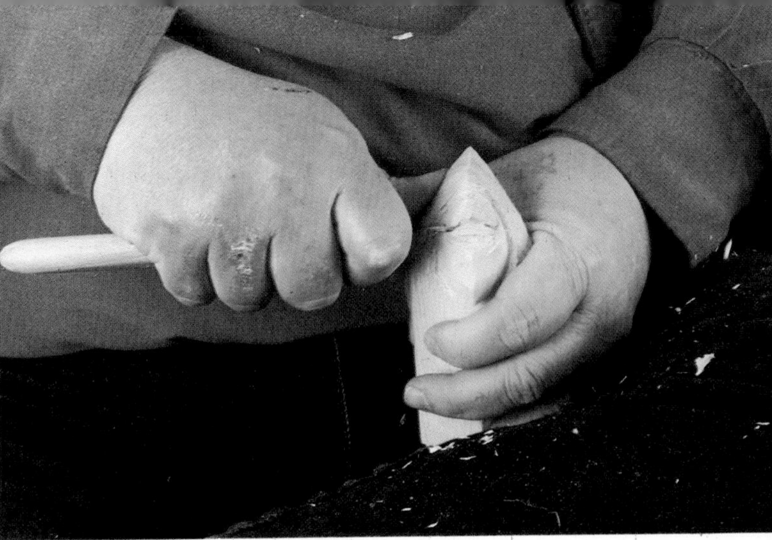

Cut a stop in the top line of the moustache.

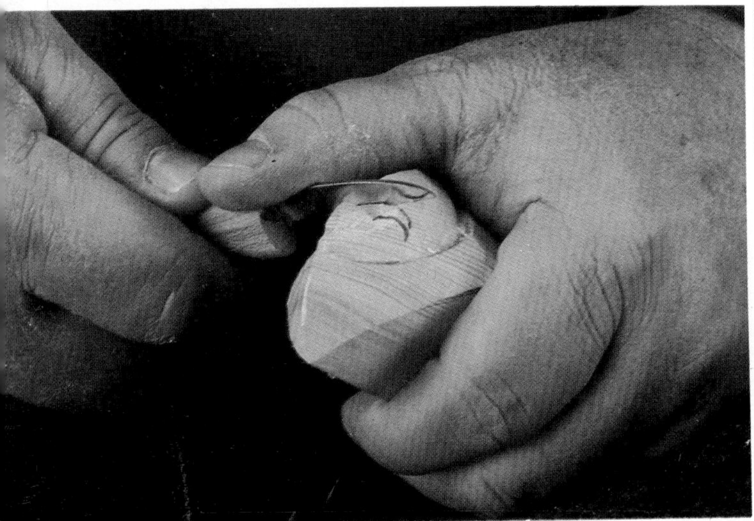

Cut a stop in the bottom line of the moustache.

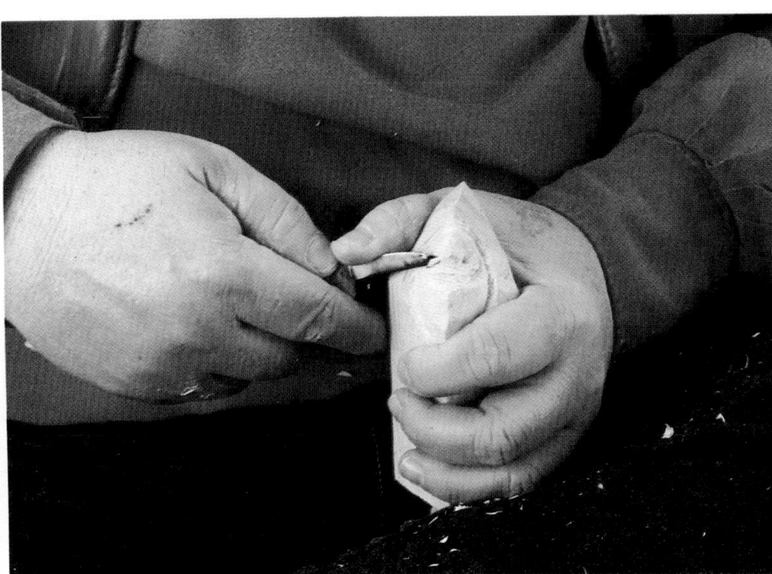

Cut back into the stop creating the rounded cheek and the shape of the moustache at the same time.

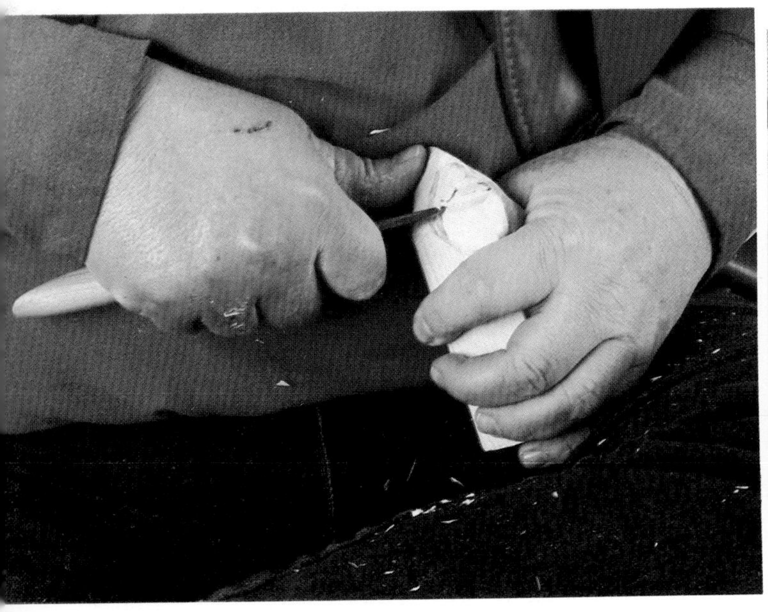

Cut back to the stop all around.

Carry the line at the end of the moustache up the side of the face to define the sideburn.

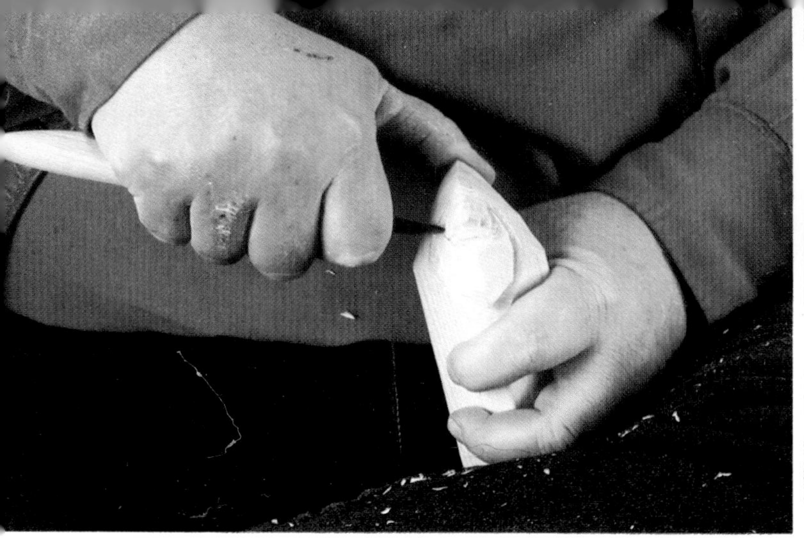

Cut a stop and carve to define the cheek and the sideburn.

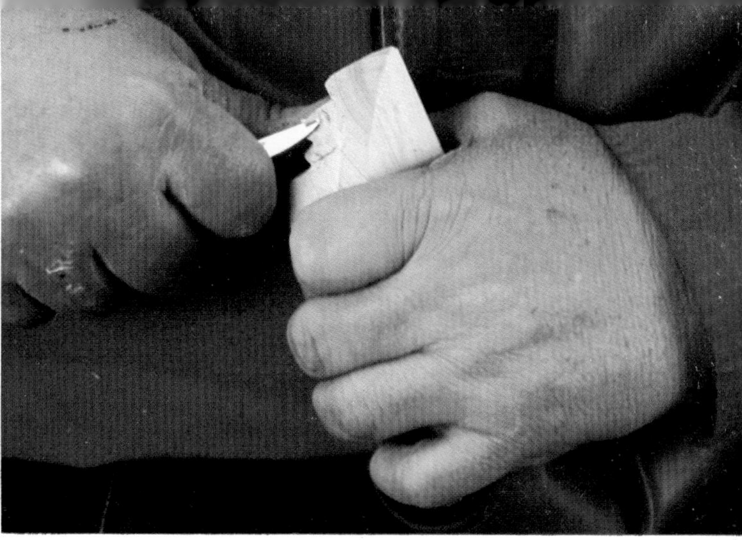

Define the line of the bridge of the nose.

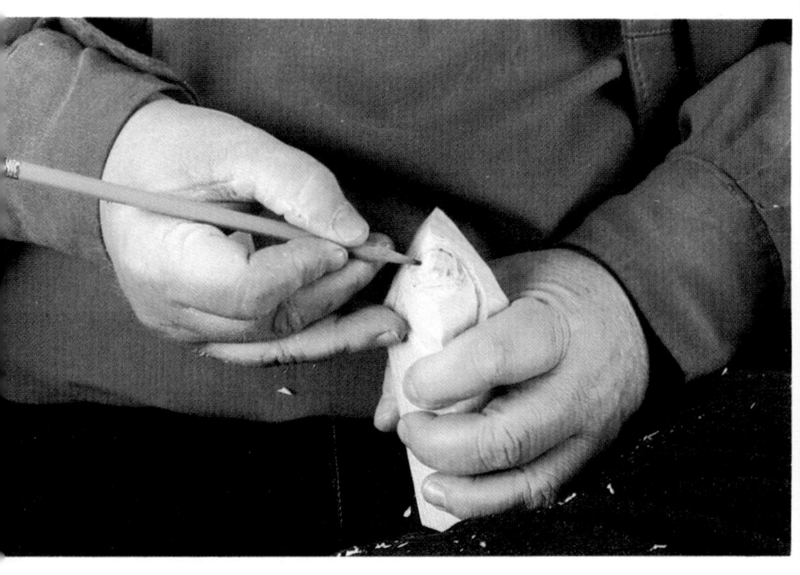

Mark a line at the eyebrow.

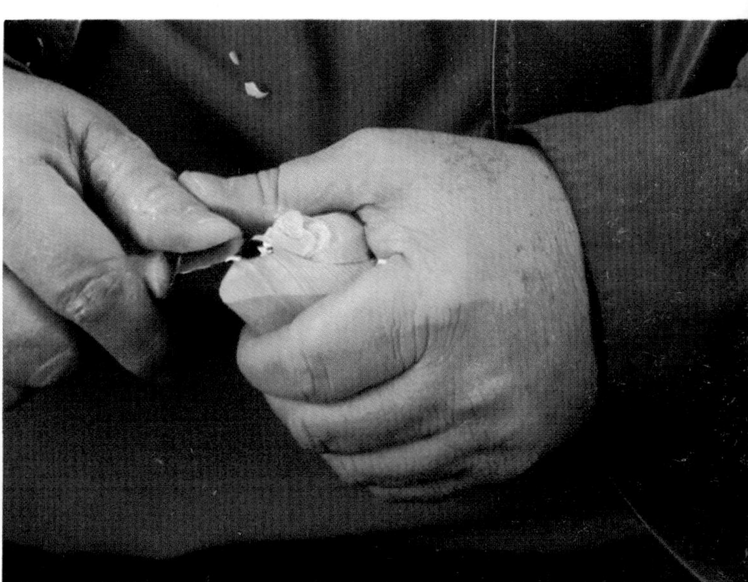

Clean up the carving...

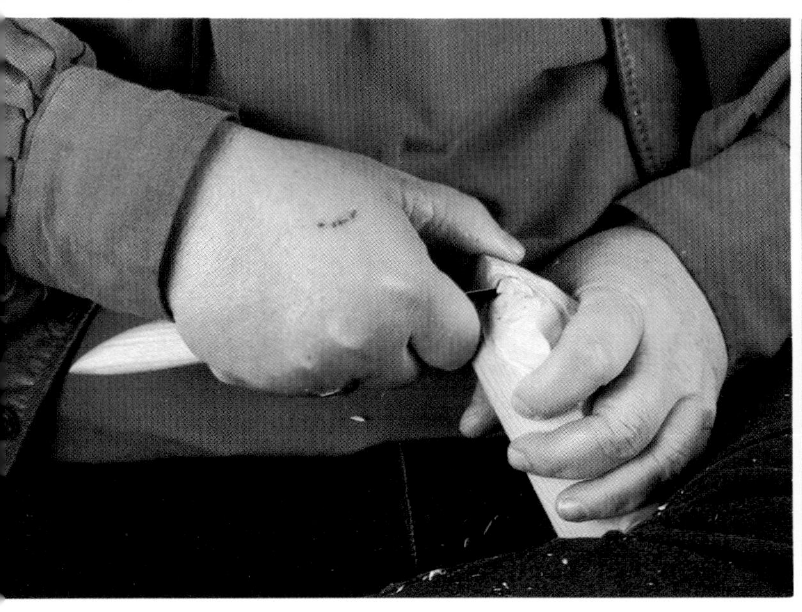

Cut out the eyesocket.

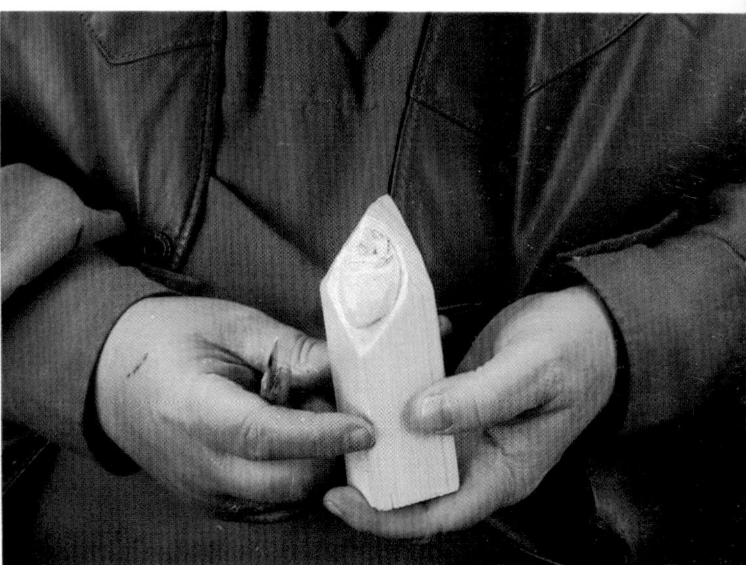

and round off the edges, removing the "hatchet" look of the face.

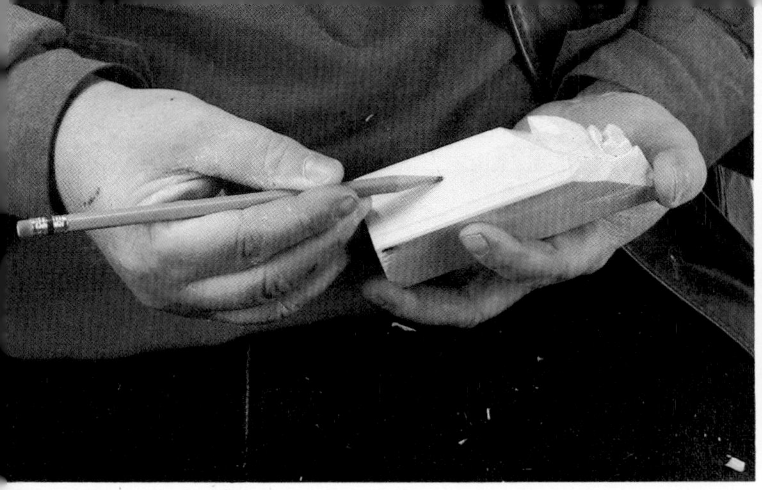

Mark the bottom of the coat. It should be more than halfway down the figure. When you find the spot, lock your fingers together and use them as a depth gauge. I often make these figures in floor-length robes, as you will see in the examples. If that is what you wish to do, move ahead.

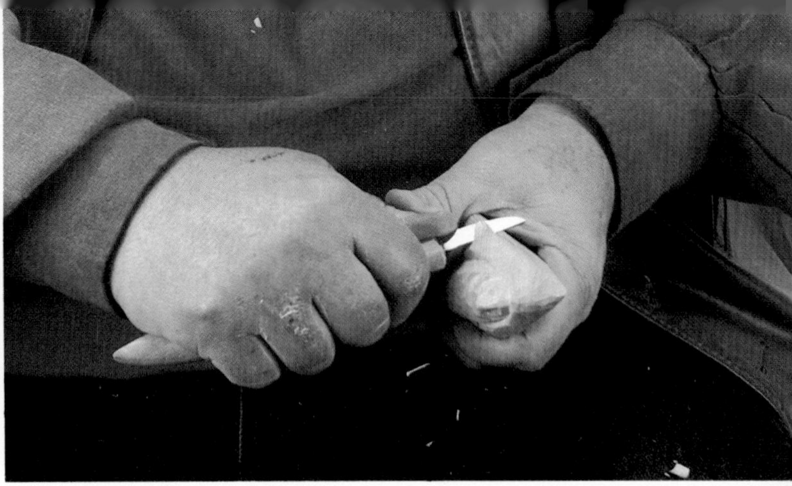

Notch out the shoulder.

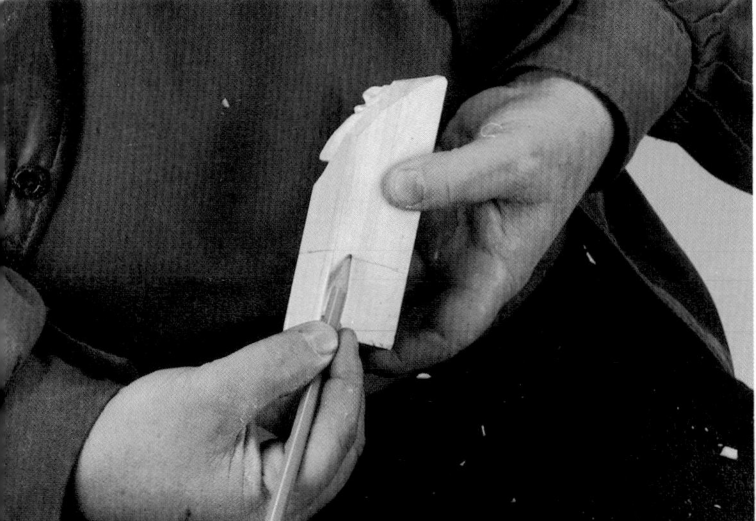

Carry the mark all around.

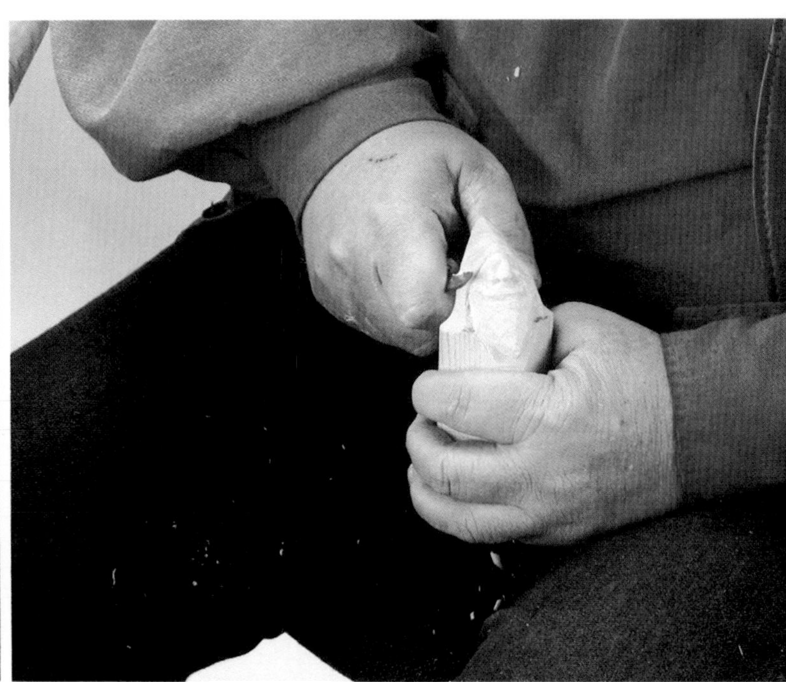

Clean away excess wood using nice smooth cuts. This will make it finishing easier.

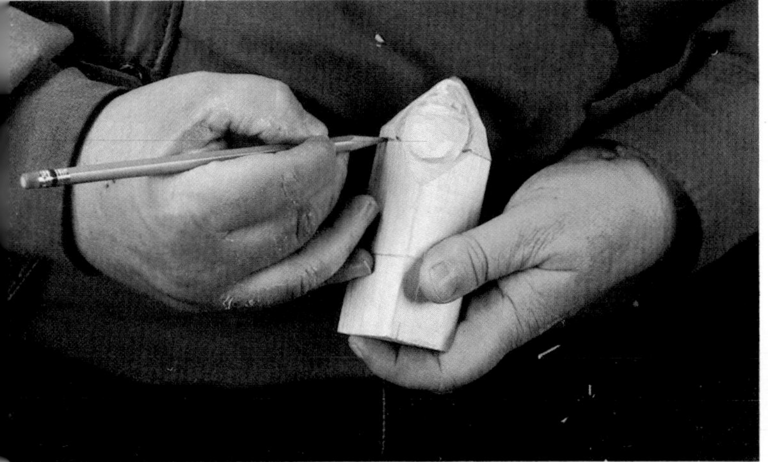

Decide where the shoulders are to be. Don't make the figure hump shouldered. Visualize where they belong, and mark the front edge.

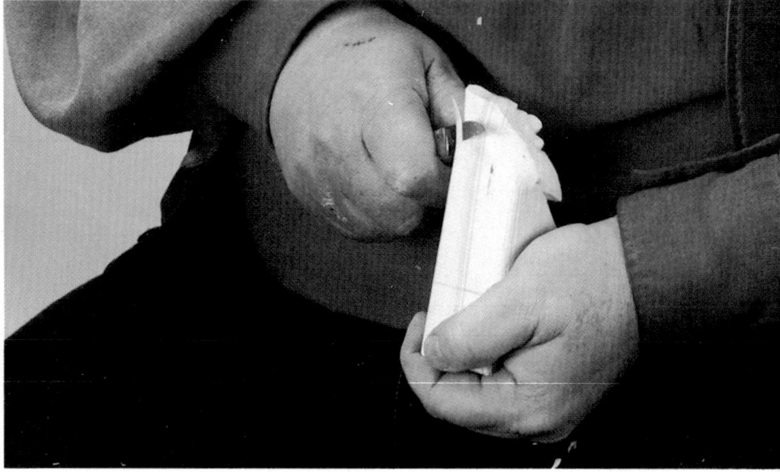

Put an arch in the back. Start at the point of the hood and cut down. This protects the point.

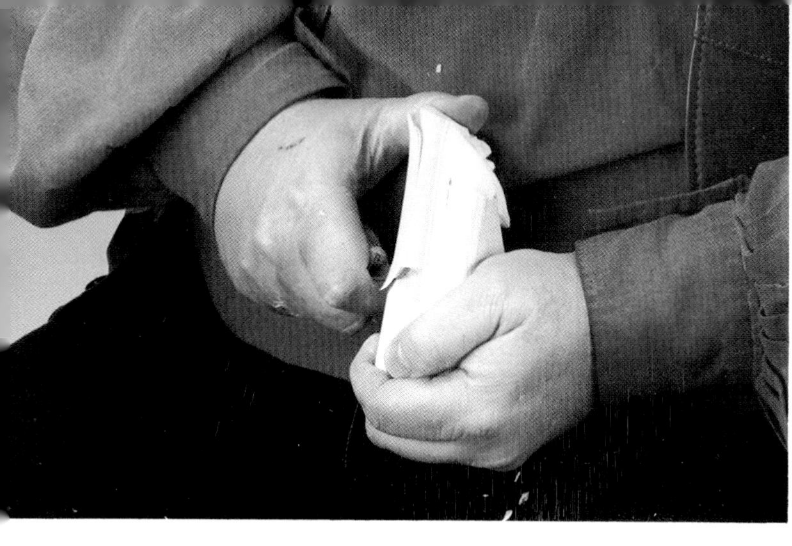

Also begin a cut at the waist and cut up. This prevents splitting.

Decide on the position and length of the arms. Often these Santas have their hands clasped in front. The curve of the back gives this Santa a forward-moving look, of which we will take advantage. We'll put him on skis, with his arms reaching forward to hold ski poles. Never let the hands stop at a coat line. Either make them longer or shorter. We want the coat to appear long so we make the arms shorter.

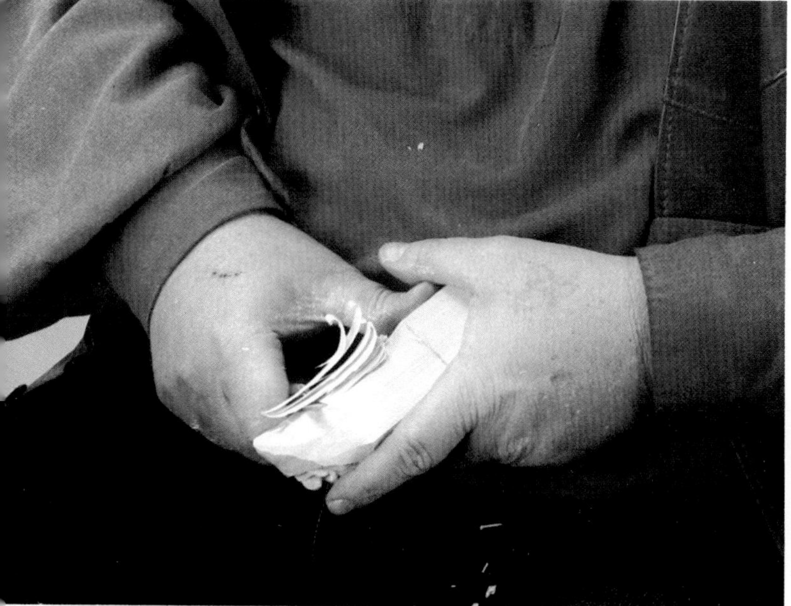

Remove the excess material by cutting down...

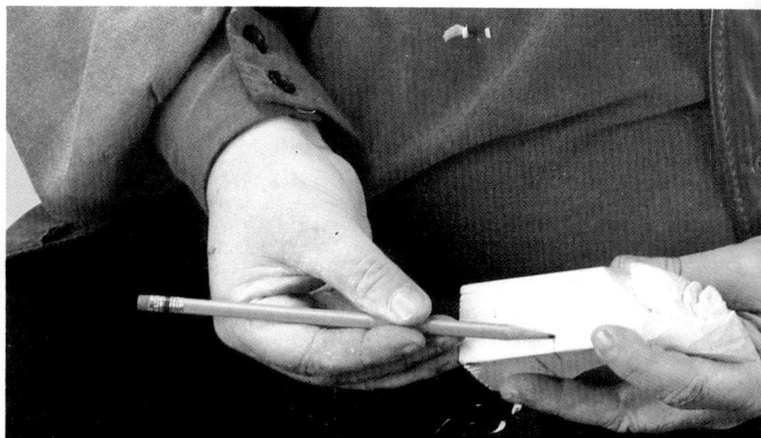

Again, use your fingers as a depth gauge.

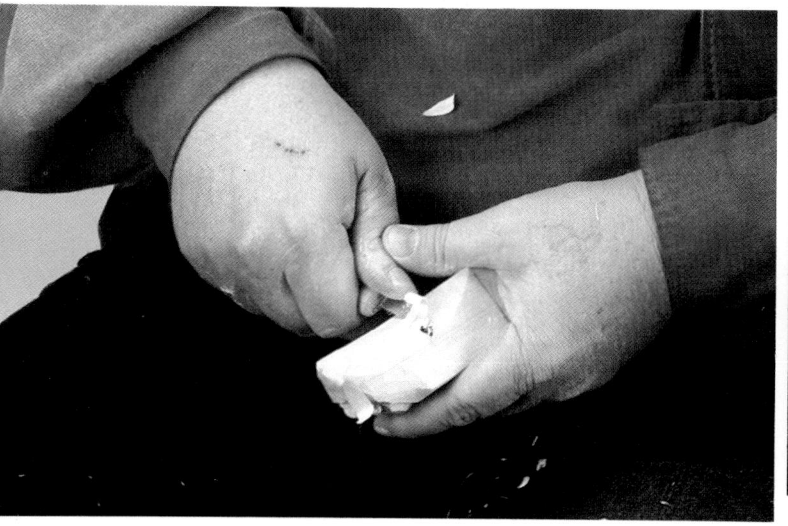

and flattening the back.

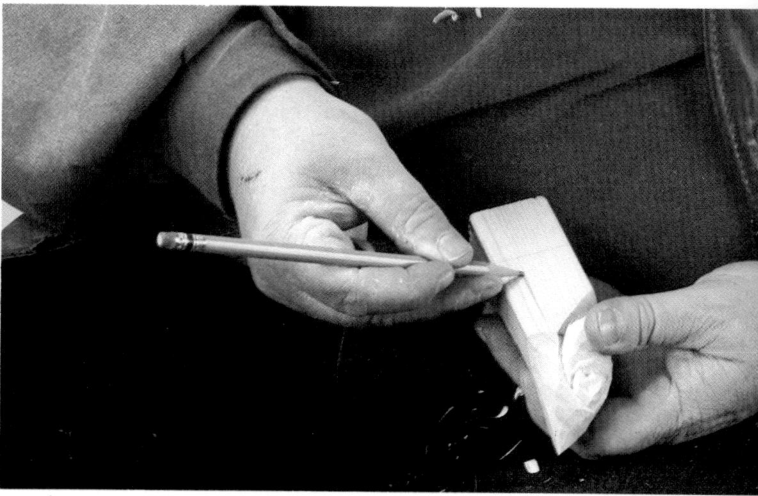

Use the finger depth gauge to mark the thickness of the arm.

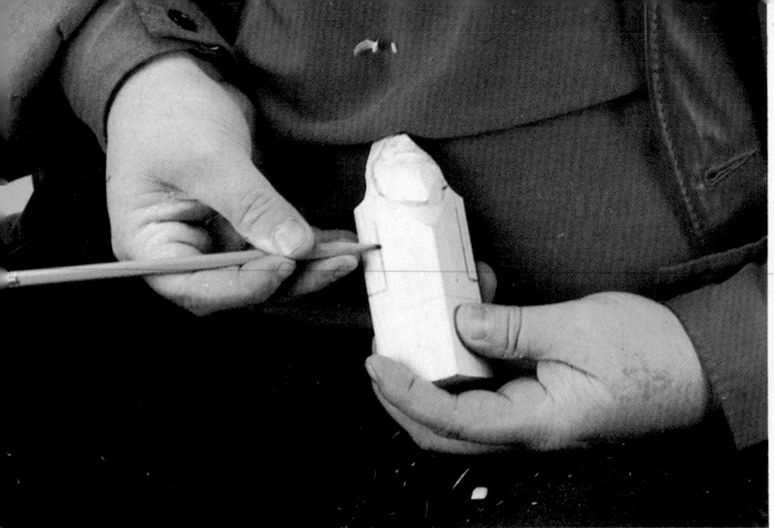

Do the other side as well.

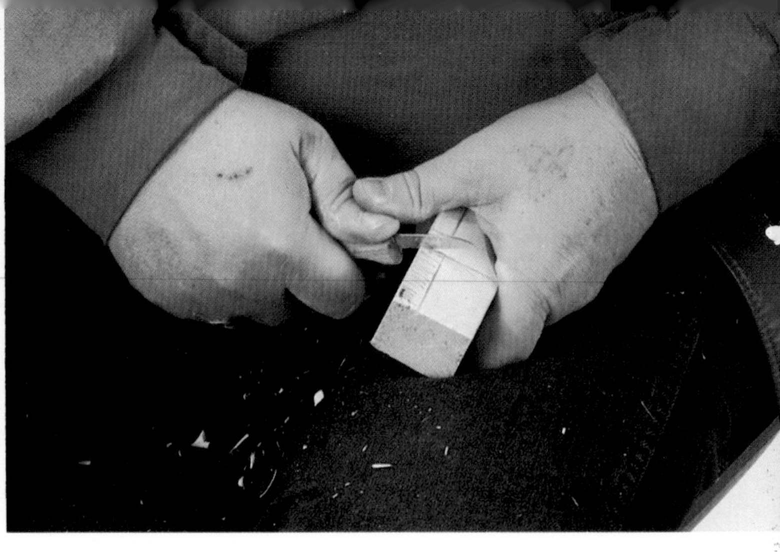

Remove the excess wood in the leg area. First, be sure to cut a stop.

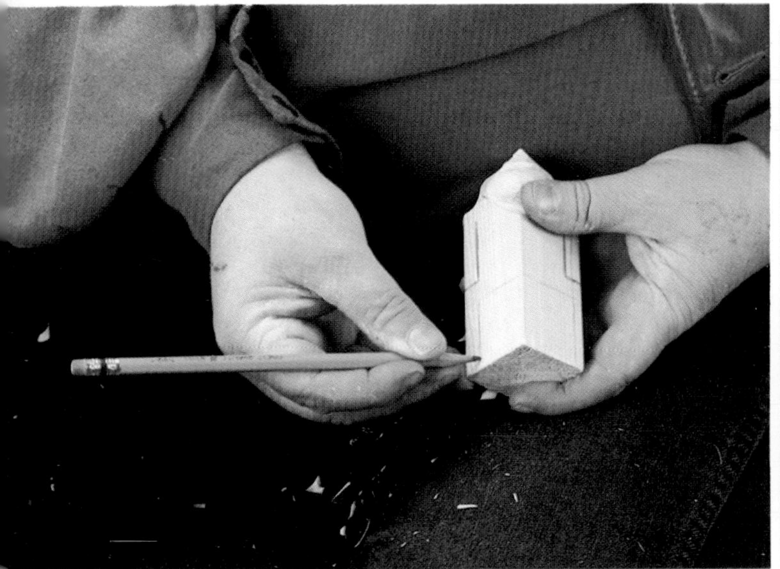

Mark the line of the legs. Following the arms' thickness usually works well.

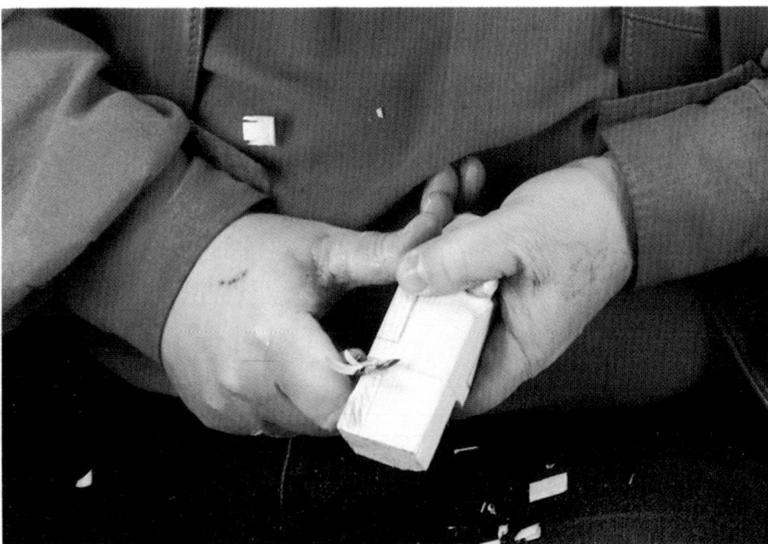

Then cut back to it.

Mark the areas to be removed.

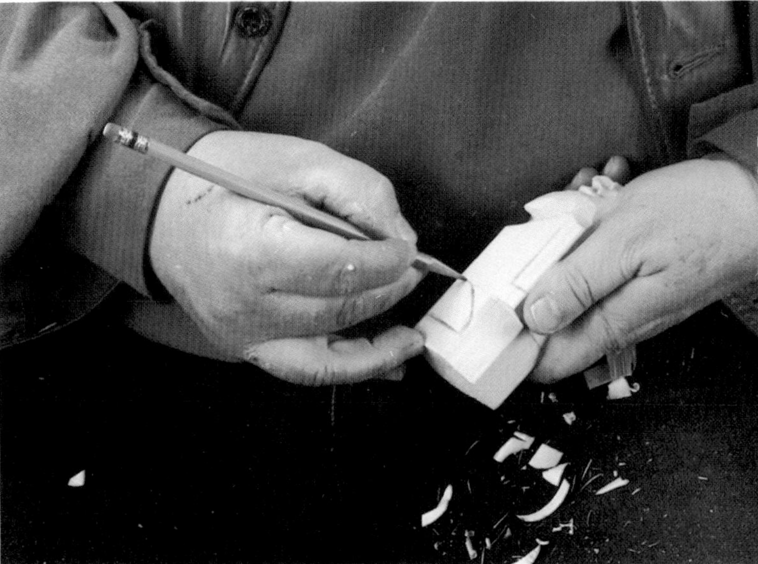

Mark the feet. I usually make them oversized to be sure the figure will stand.

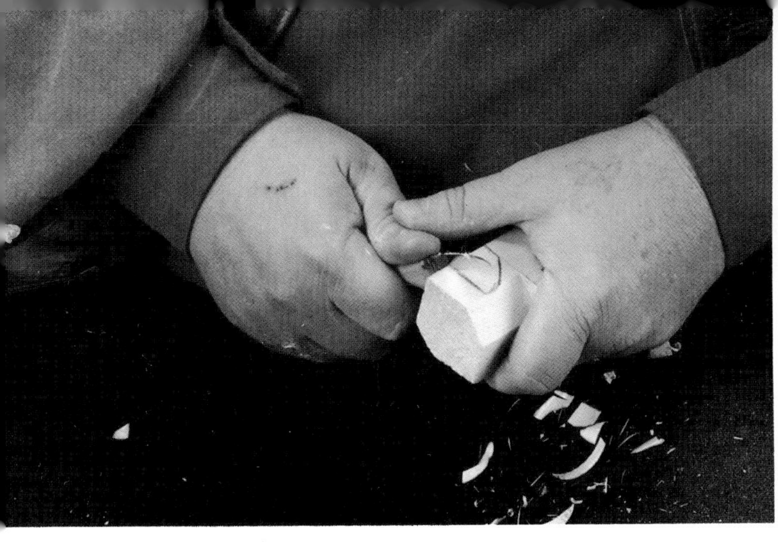

Cut a stop at the top of the feet...

It is beginning to look like a Santa.

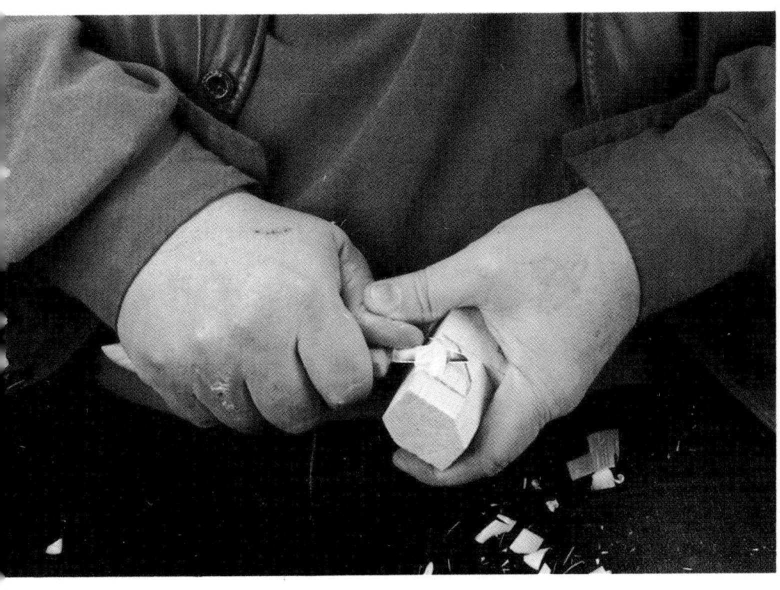

and cut down to it.

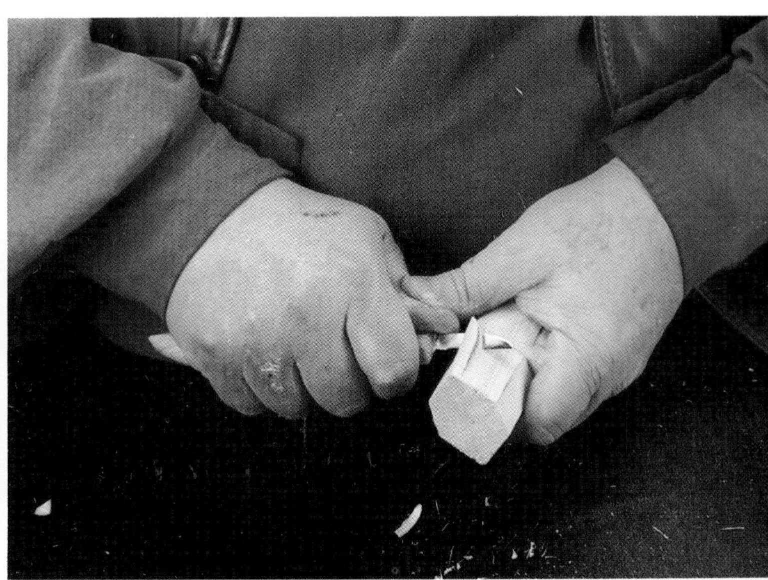

Start at the bottom of the coat...

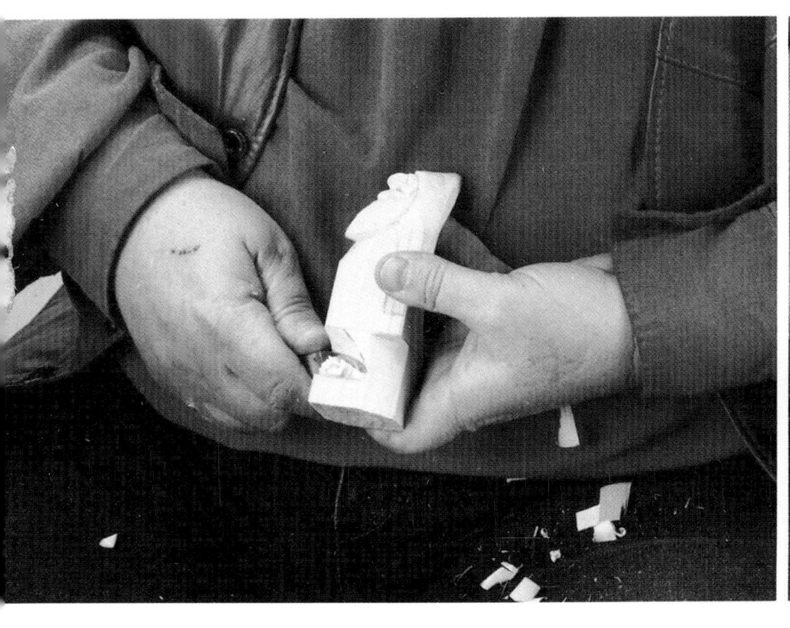

Smooth out the area.

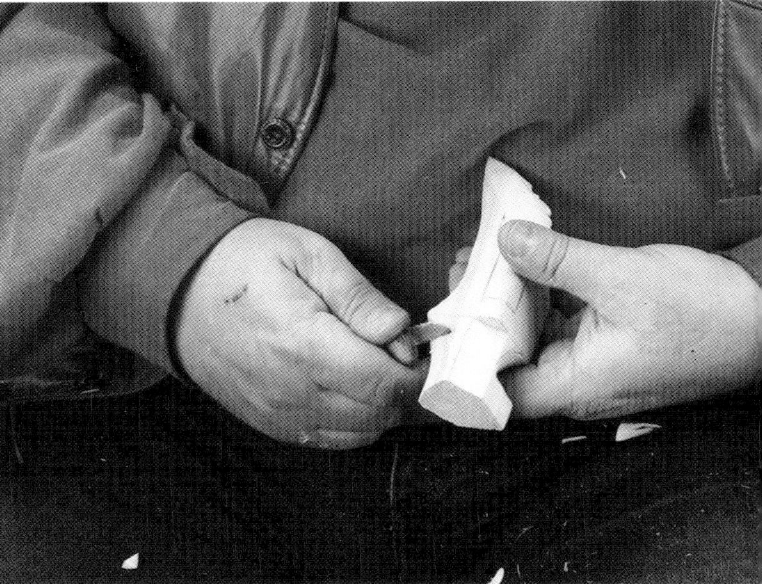

and trim down to the heels of Santa.

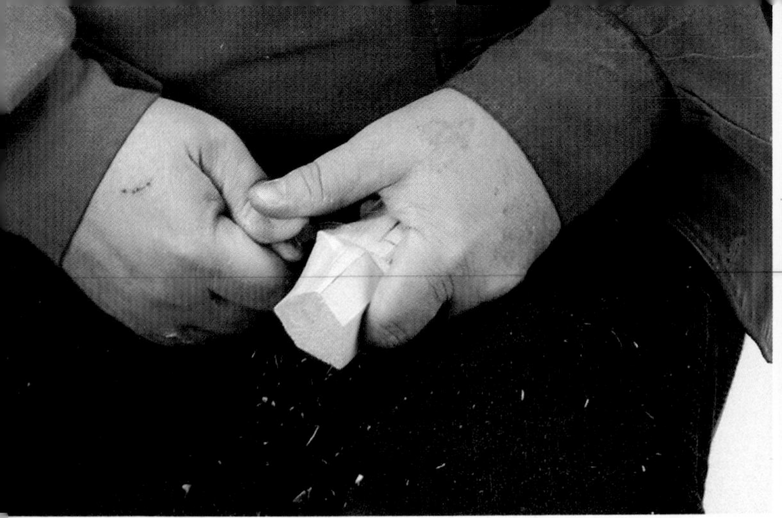

Cut a stop at the back hem of Santa's coat and cut back to it.

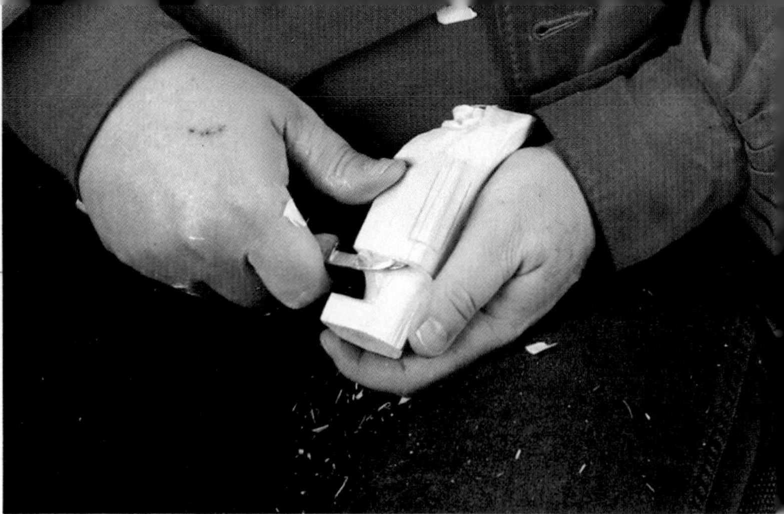

and cut back to it to bring out the coat line.

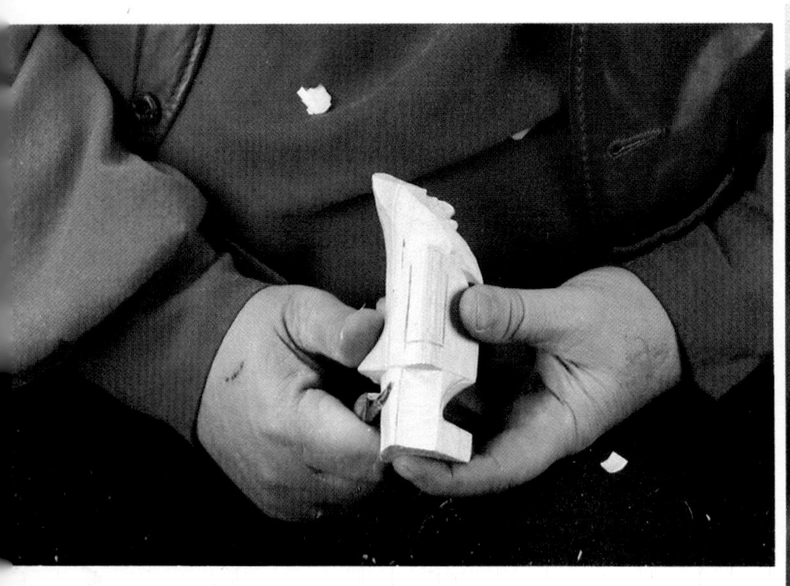

This brings the line of the legs up.

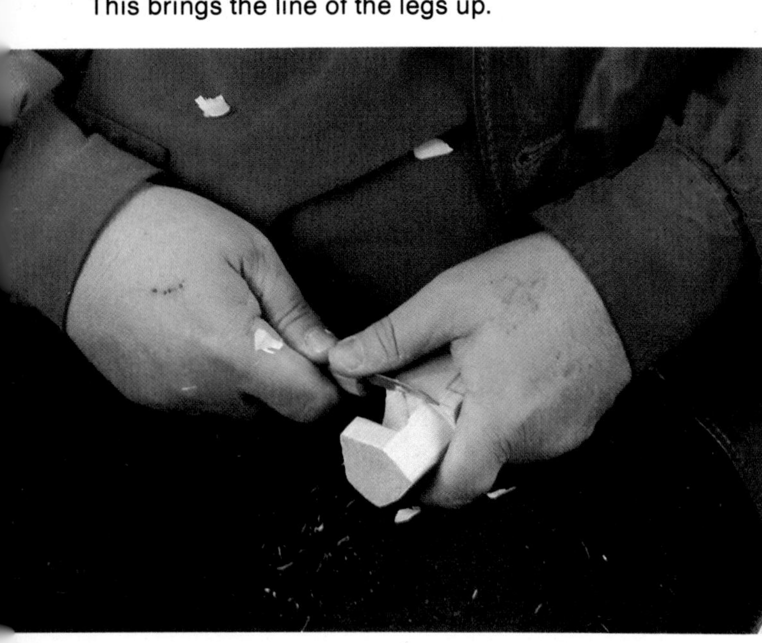

Follow the hem of the coat all around with a stop line...

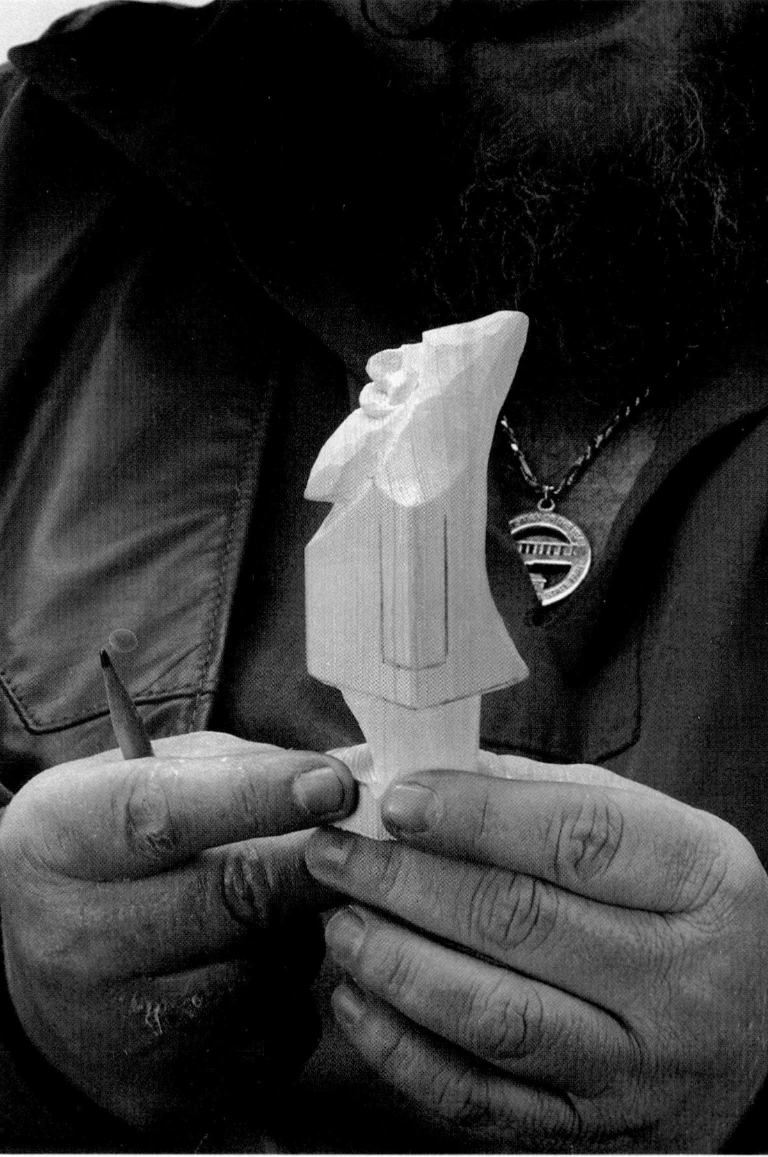

You are now at this point. While we have decided on a skiing Santa, it should be noted that we could have taken out the curve in the back and made him more traditional. One of the fun things is, however, to go with the wood.

Mark the middle of the leg...

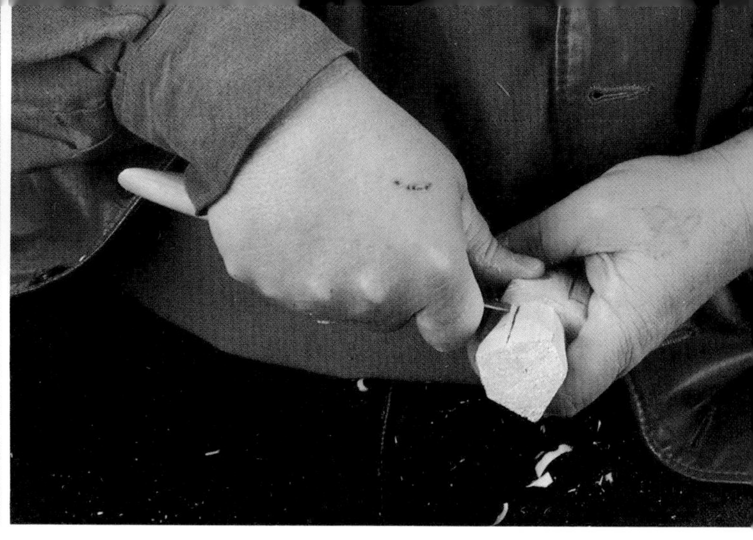

Cut a v-shaped groove on the leg line. Start on the back and cut one way first...

and carry it through the foot.

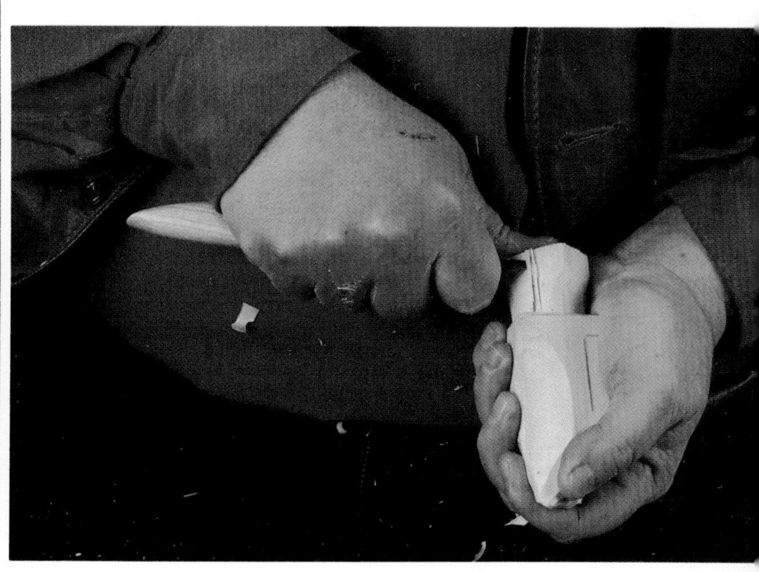

then, turning the figure, cut in the other direction.

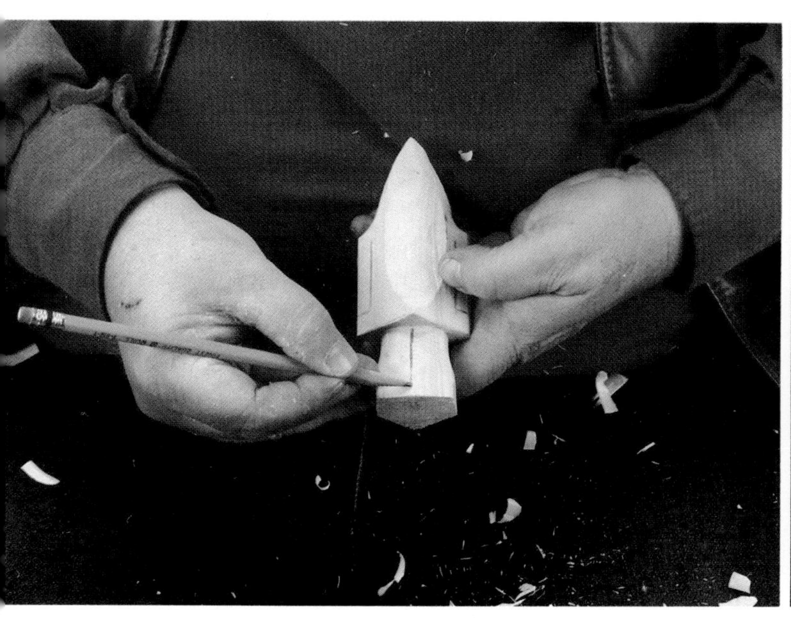

Go to the back and do the same thing.

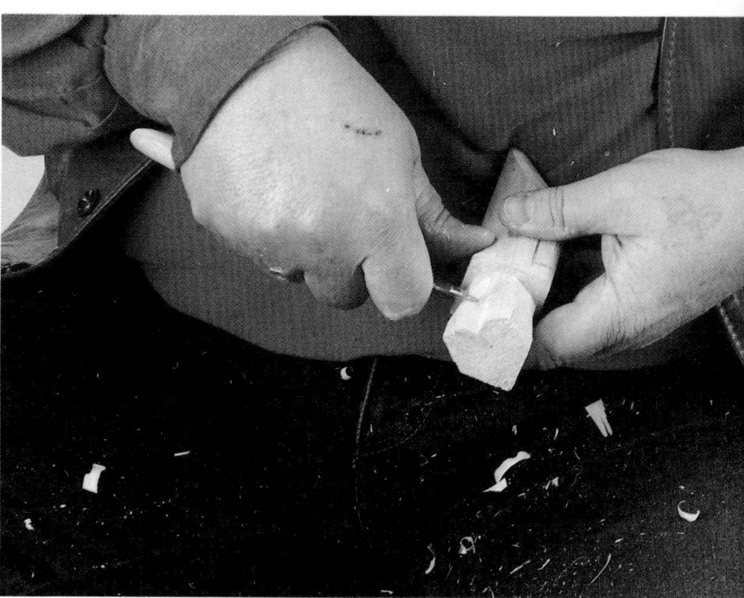

Continue widening the "v" until the definition is clear.

Do the same thing on the front

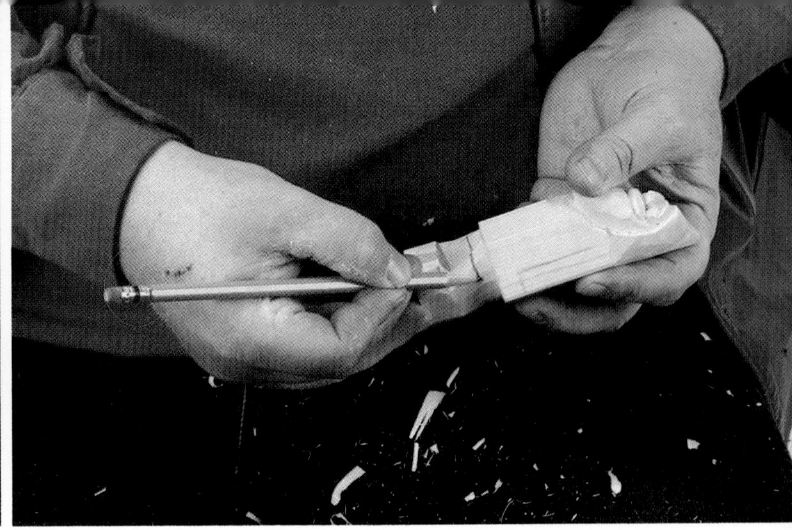

Decide how far up the leg the boot should go and mark all around.

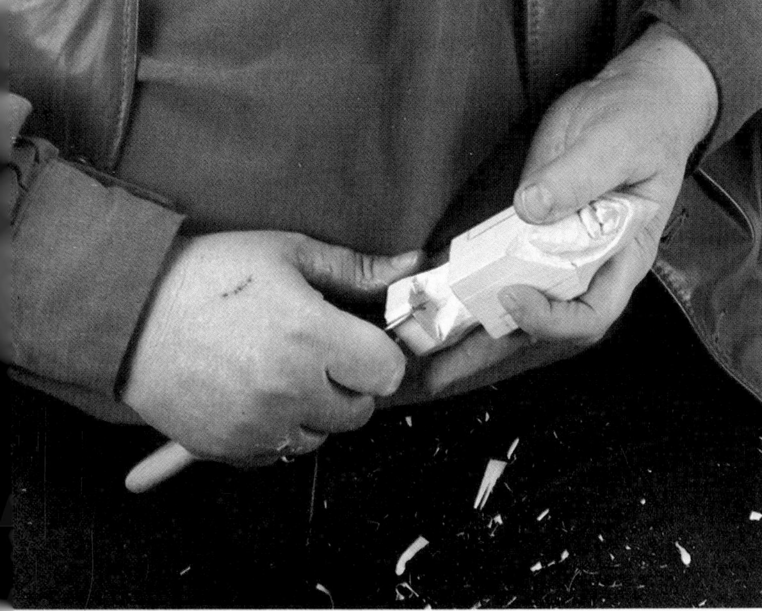

Cut a stop around the top of the boot.

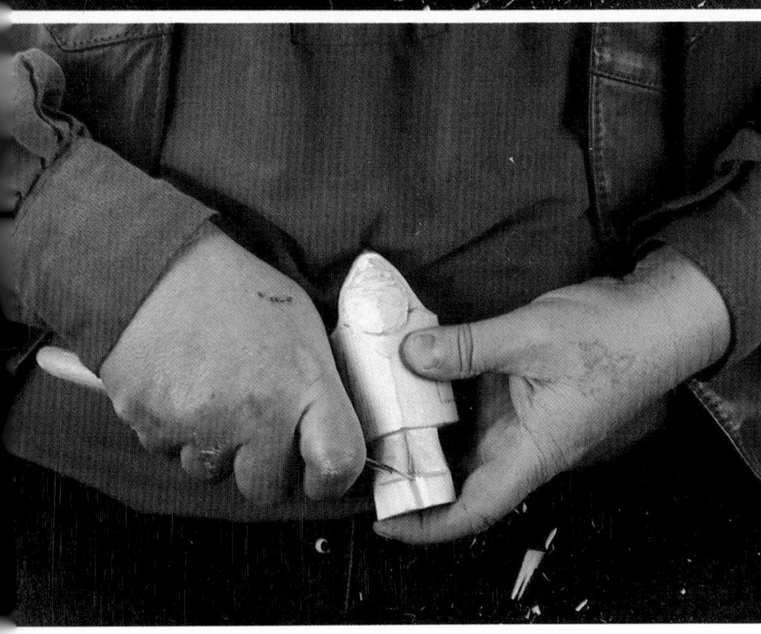

Next cut the two shoes apart.

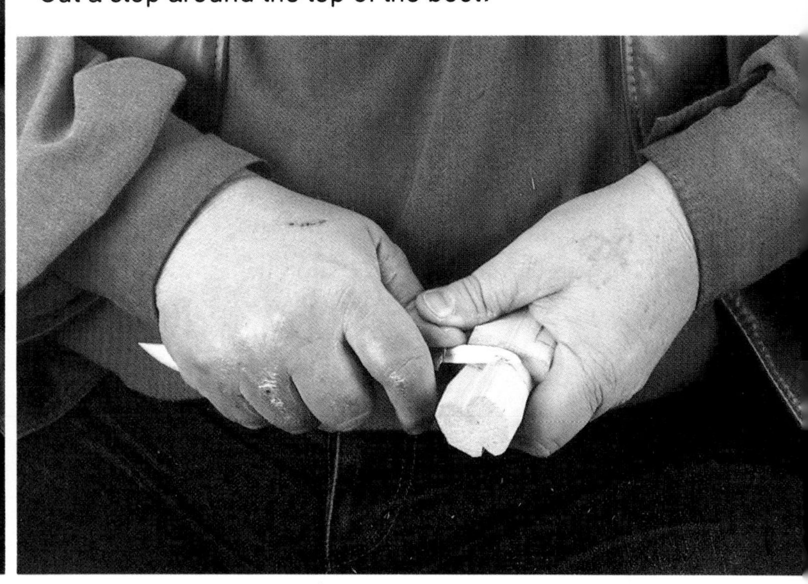

Cut from the top of the trousers down to the boot line.

Note the progress so far.

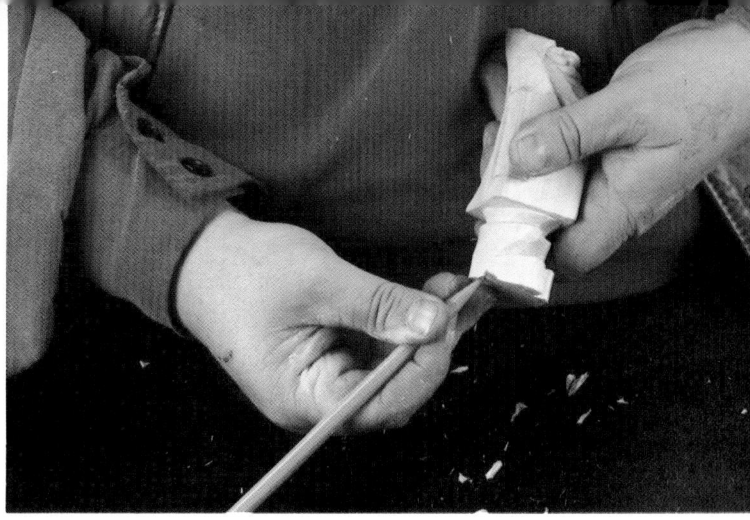

Mark a heel notch on both sides.

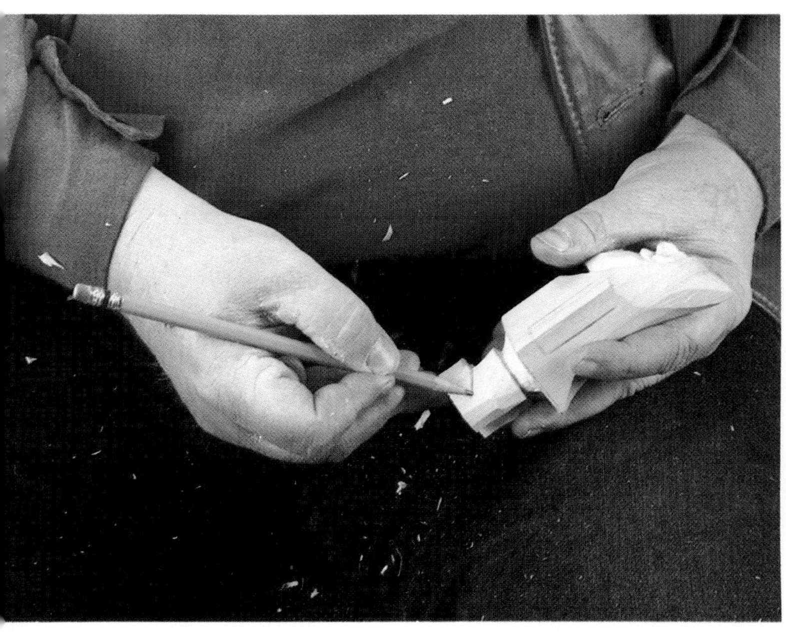

Draw character lines on the sides of the boot...

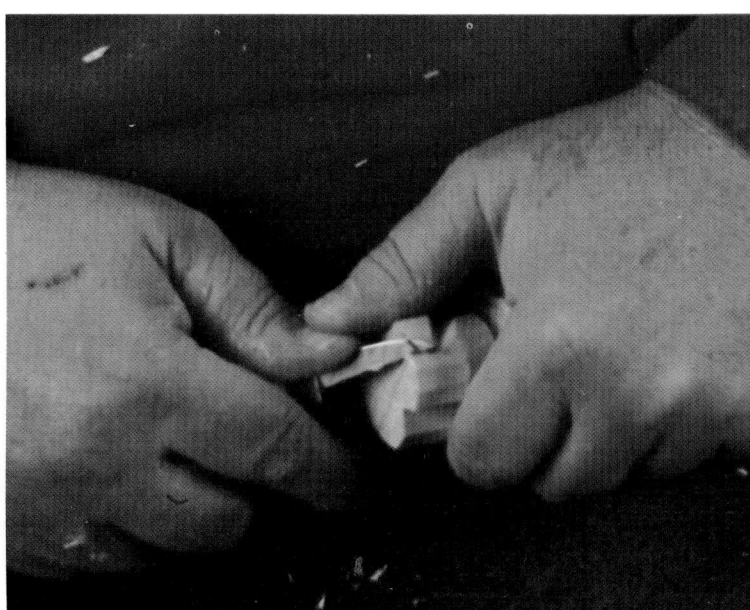

Cut a notch by carving one straight stop cut and cutting back to it.

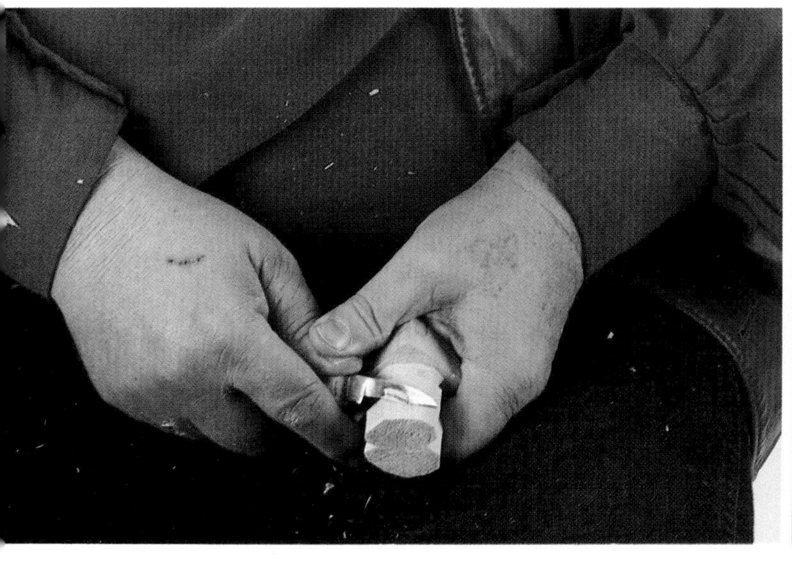

and cut a v-shaped notch on the line.

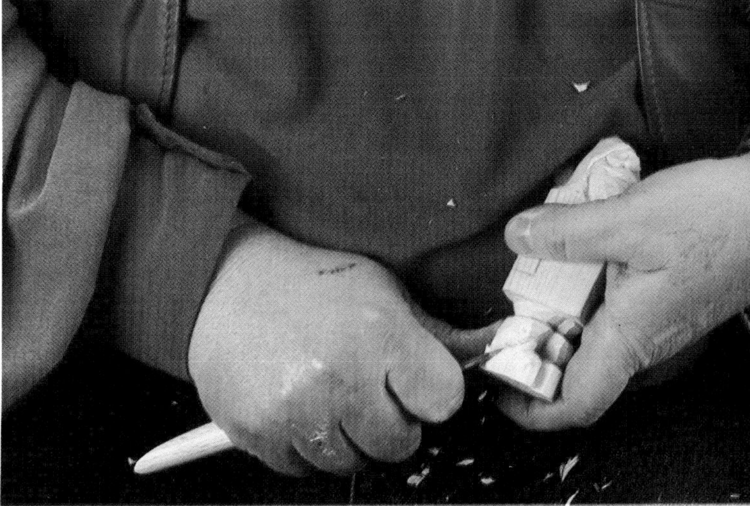

Round off the sharp edges of the boots.

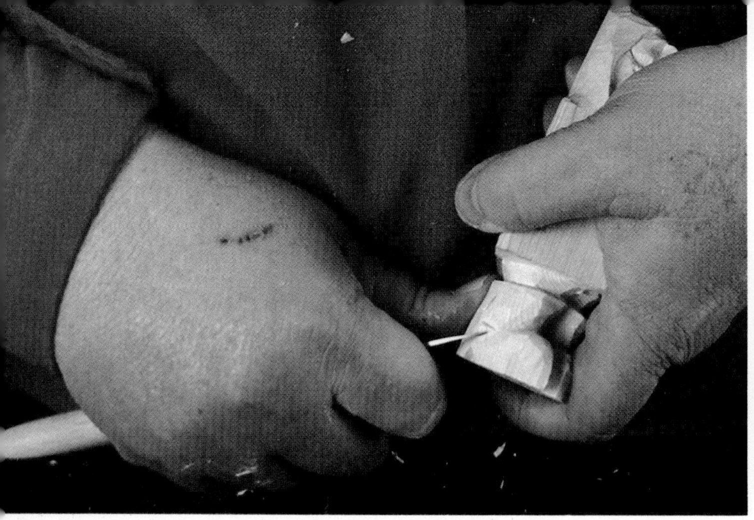

Cut three lines on the ankle to break up the boot line and give the appearance of a leather fold.

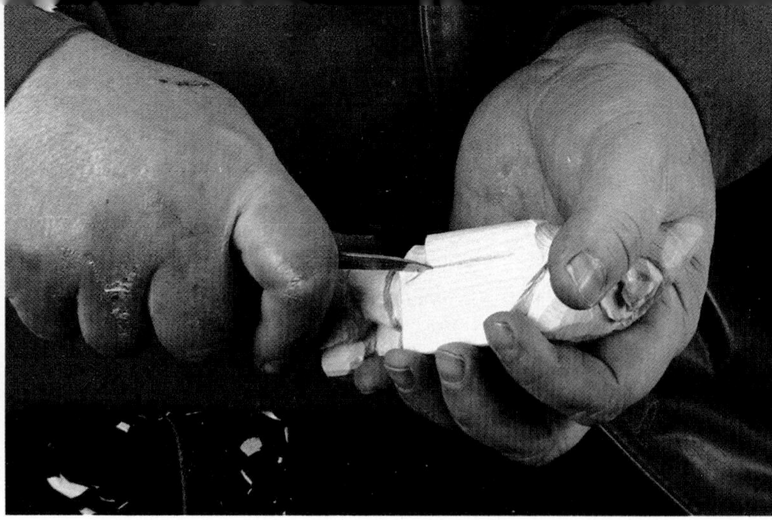

Cut a stop down the arm lines, front and back on both sides.

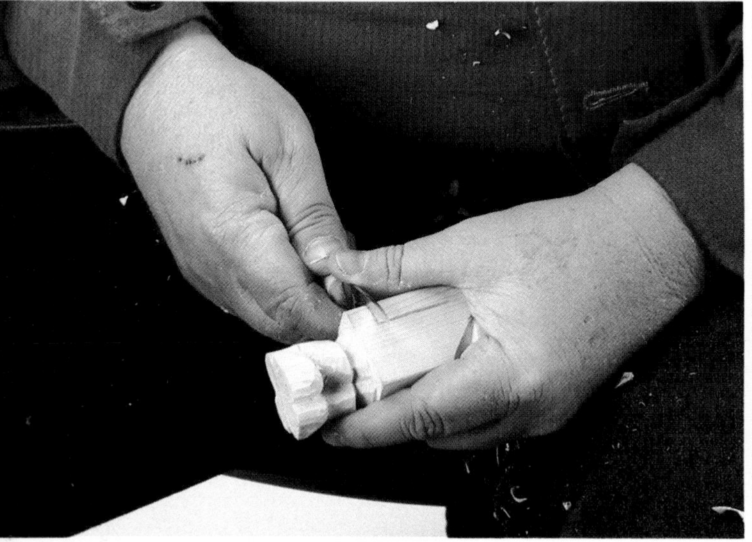

Cut a stop at the bottom of the arm.

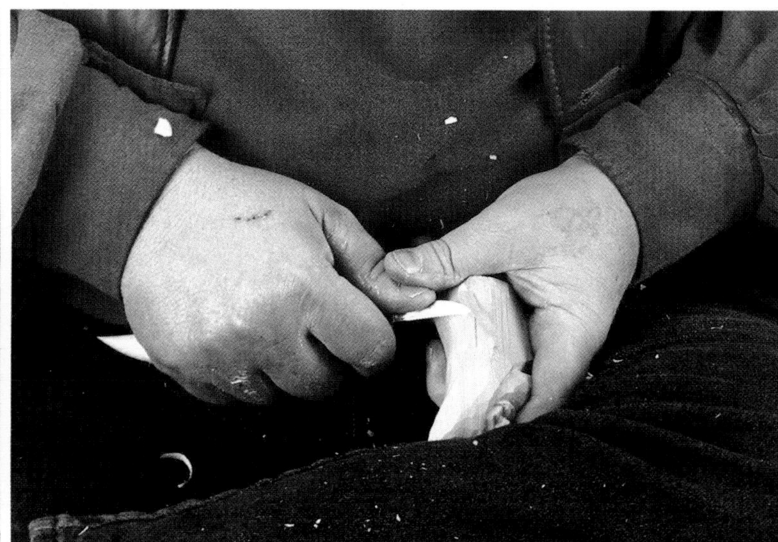

Trim away the coat to define the arm.

Trim back to it from the bottom of the coat and round the bottom edge of the coat.

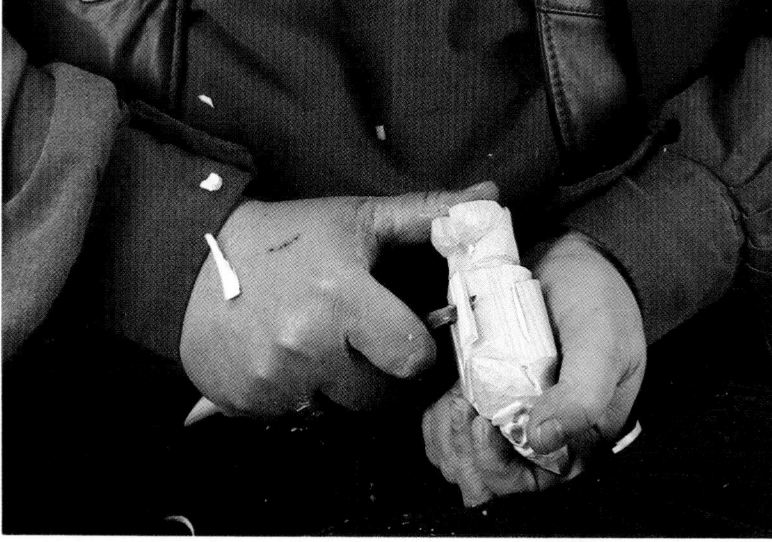

Round up the belly. Going with the grain makes it easier to remove this excess and is particularly important in larger pieces.

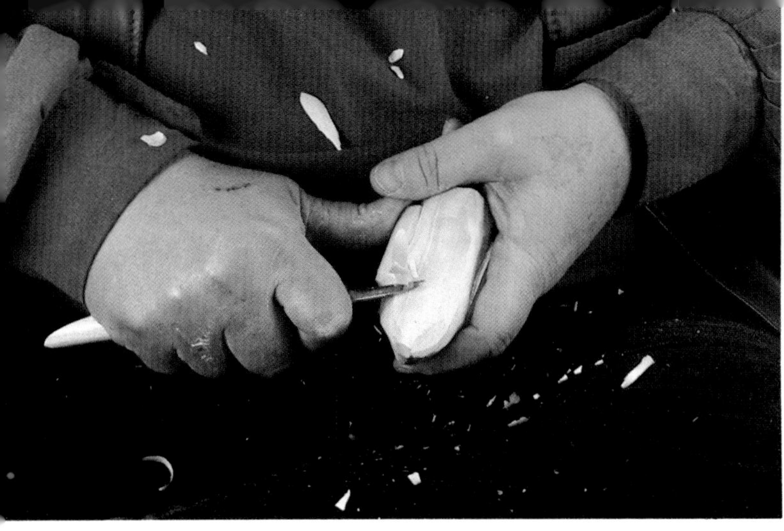

Round off everything and define the lines.

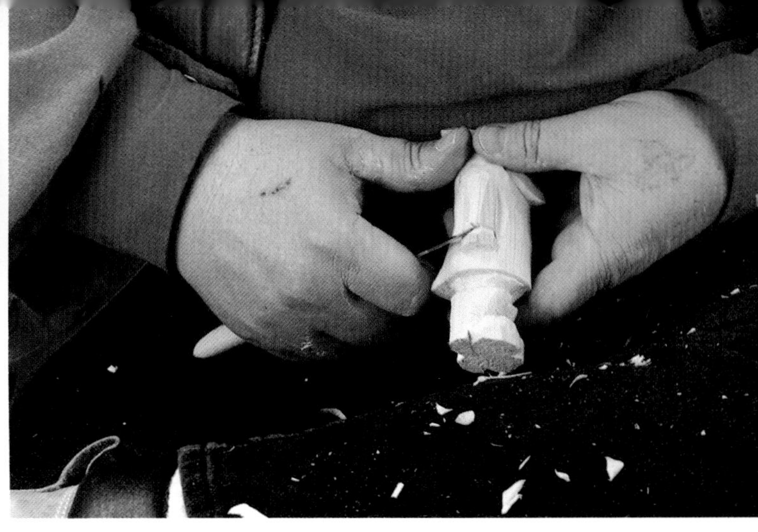

Cut back to it from the mitten.

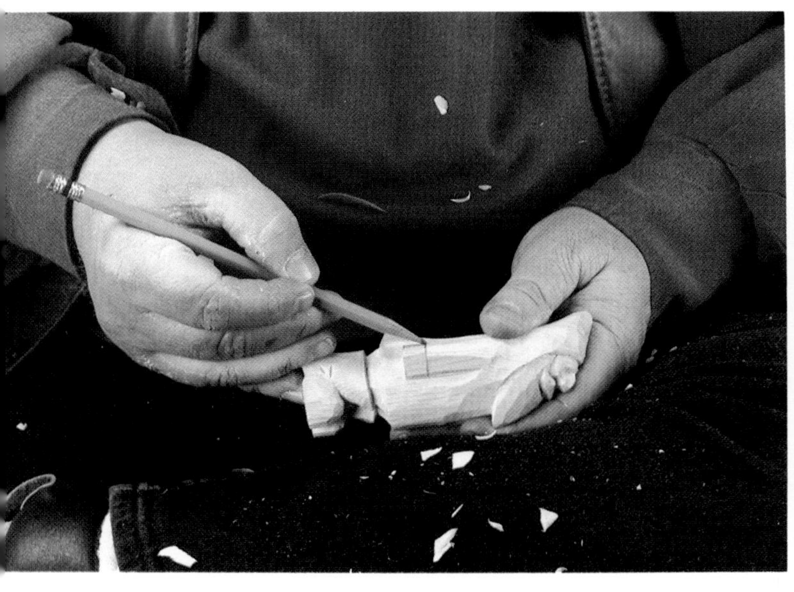

Draw the cuffs of the sleeve...

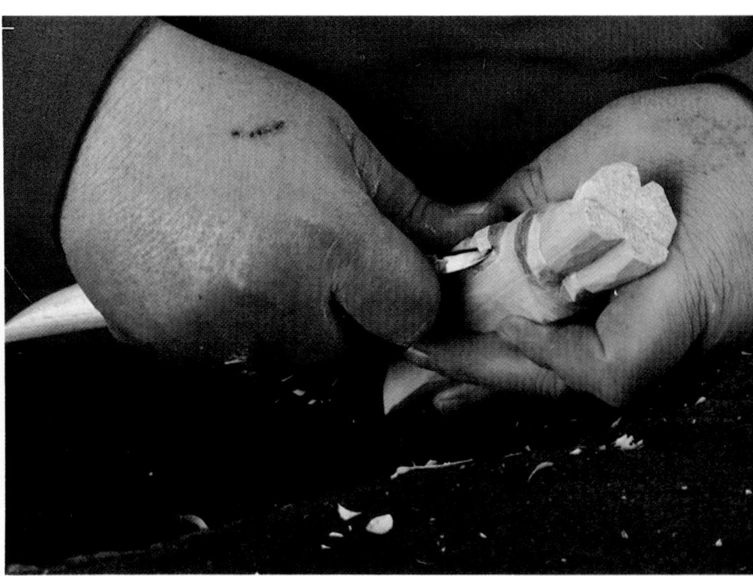

Round off the mitten.

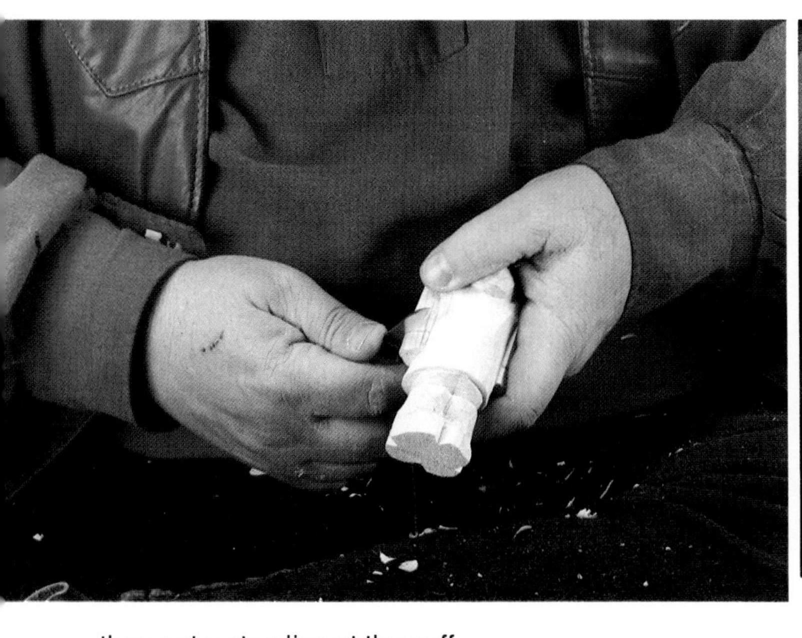

then cut a stop line at the cuff.

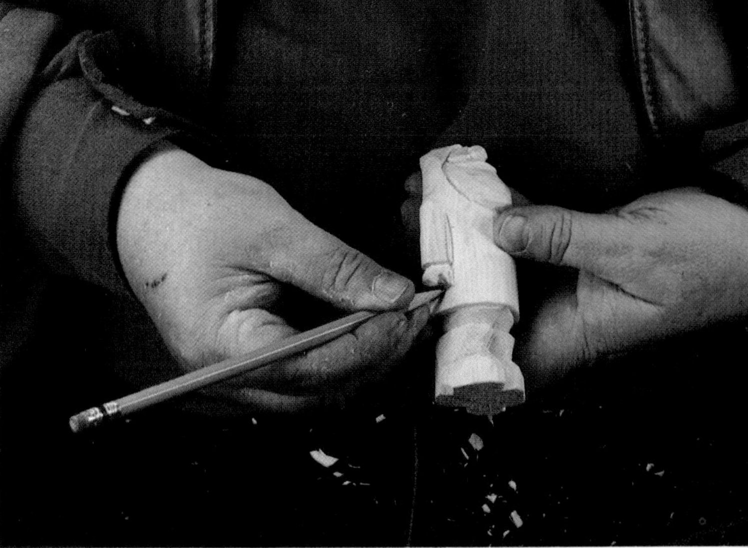

Draw and cut a small notch to define the thumb and its separation from the four fingers of the mitten.

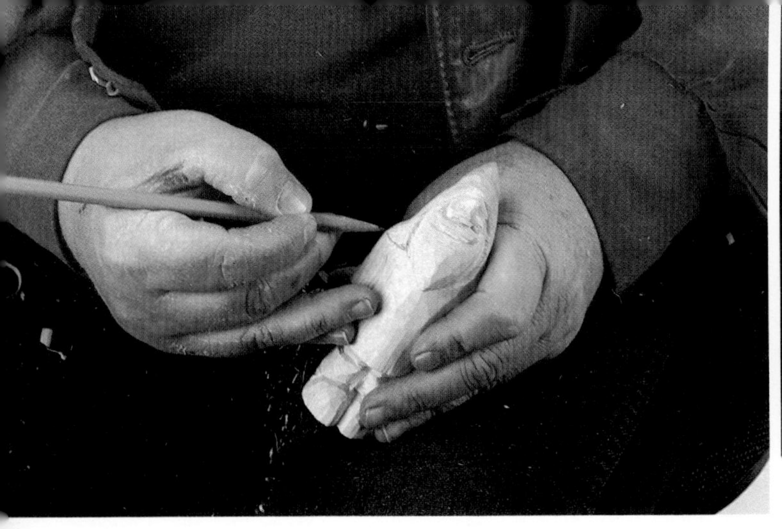

Draw the line for the bottom of Santa's hood so it begins under his beard....

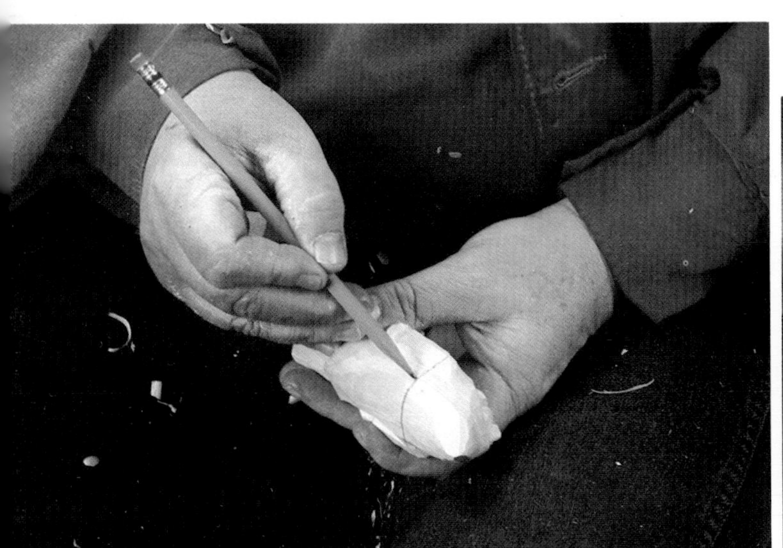

and goes around the back.

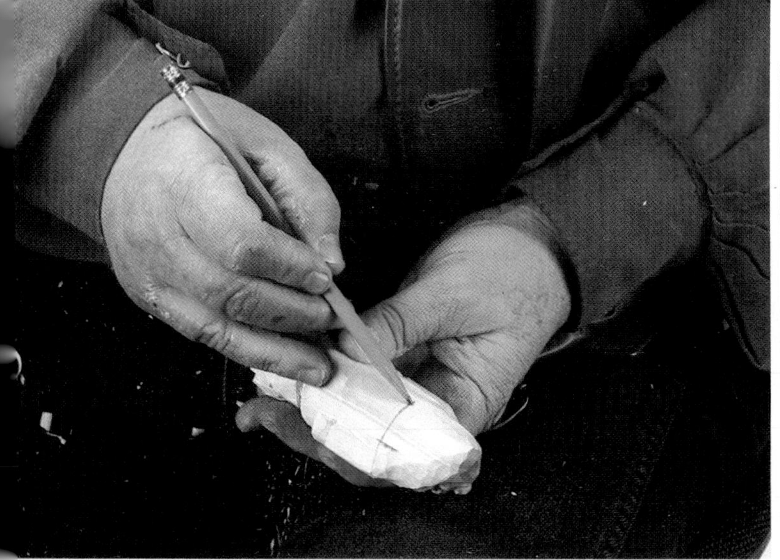

Notice that we stopped the arm line low enough so that it did not interfere with the hood.

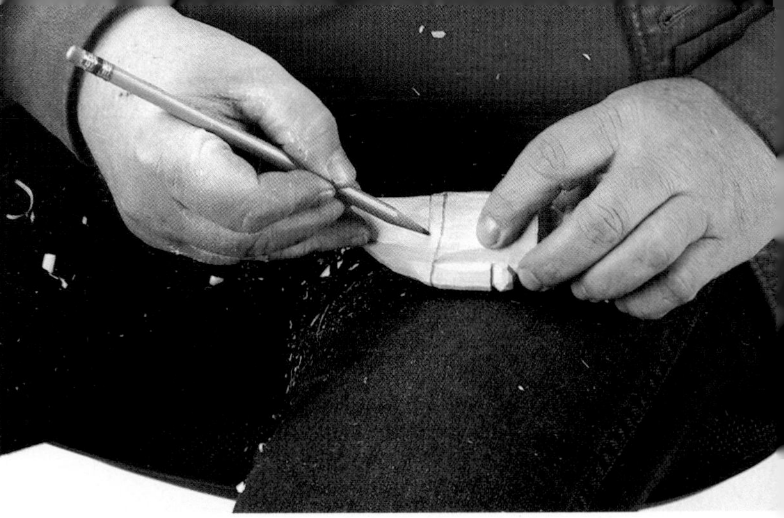

The pattern of the bottom of the hood can vary with your taste. I like a notched-out flat look with a kind of a classic Greek border. To do this I draw a second line above and parallel to the first.

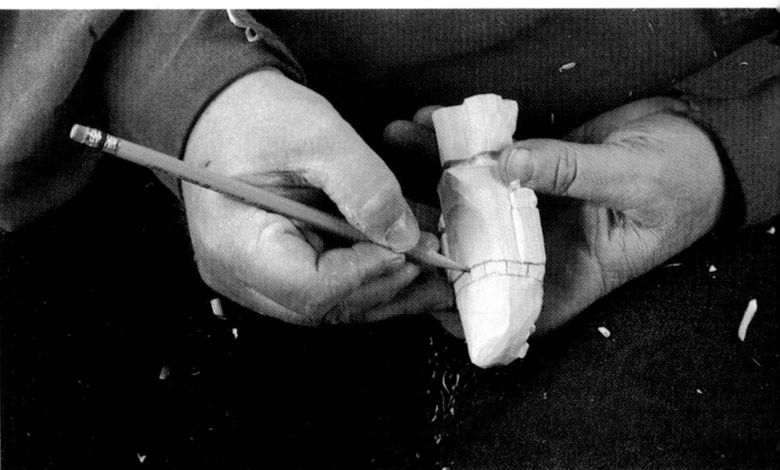

Then I divide it in sections to define the edge pattern. Count and make sure you begin on one side of the beard with the same part of the pattern as you end with on the other side.

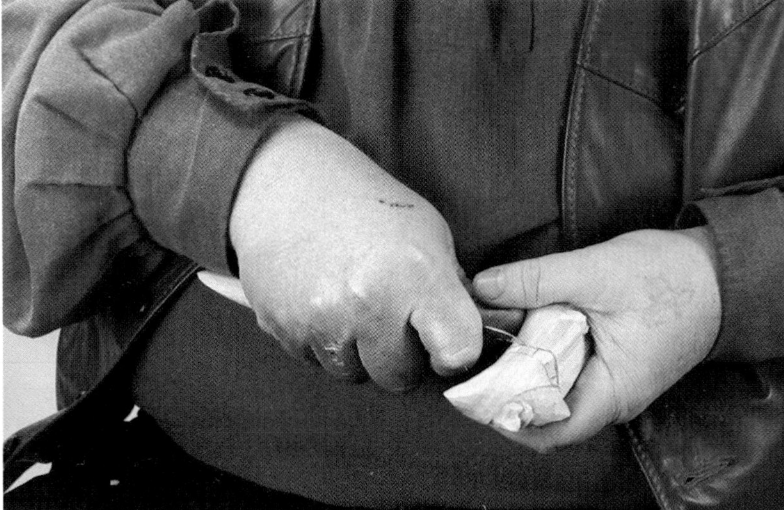

Carve a stop across the bottom of the hood.

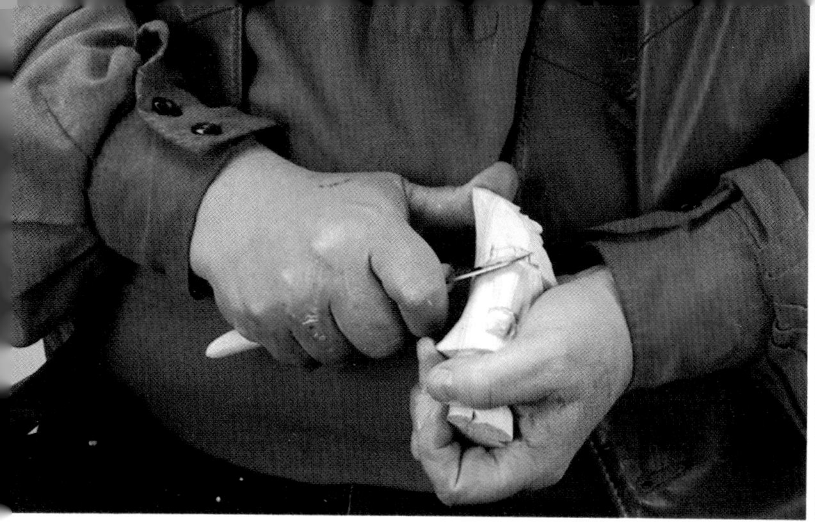

Cut up the coat to the stop.

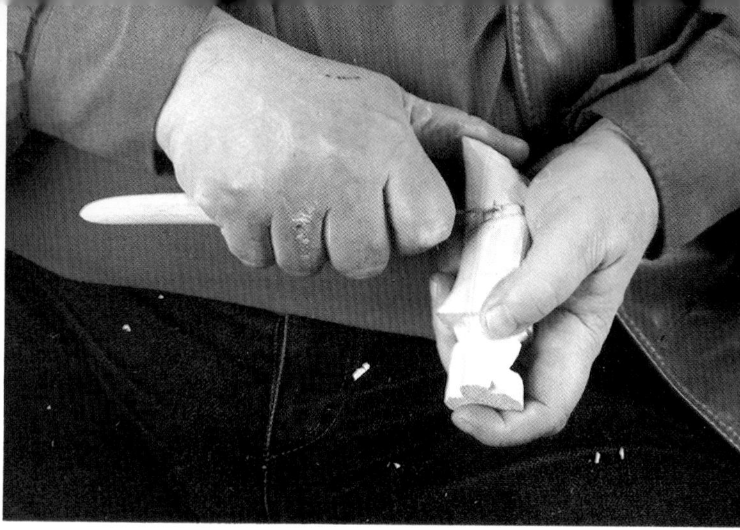

Chip carve the excess wood out of the pattern.

Go straight into the edge pattern of the hood.

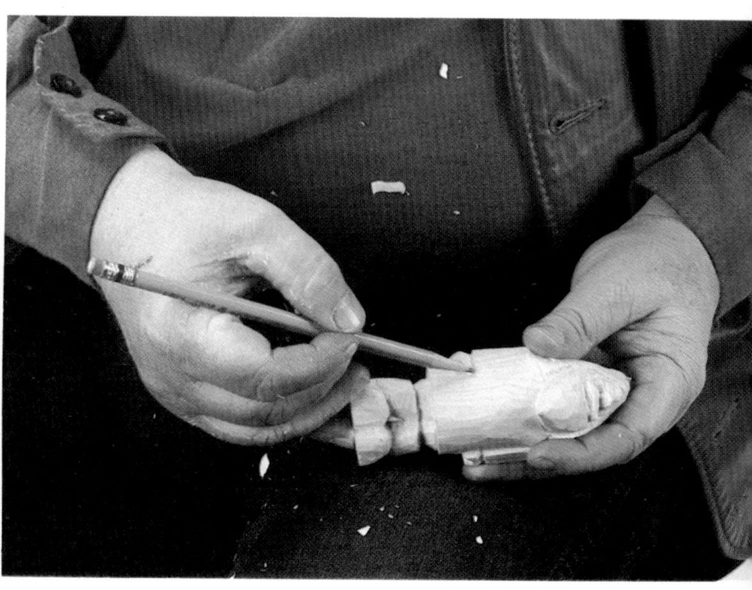

The belt is next. Don't have it meet another line.

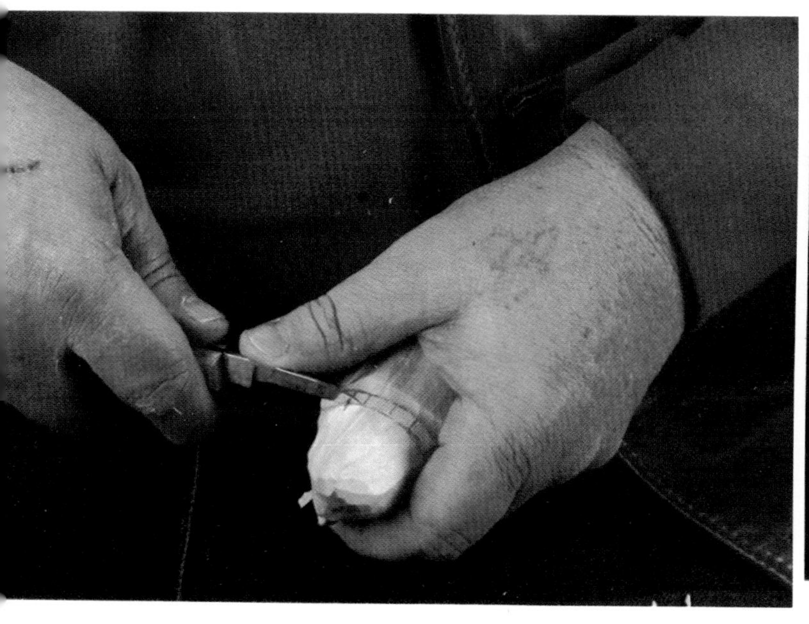

Do the same on all sides of the pattern.

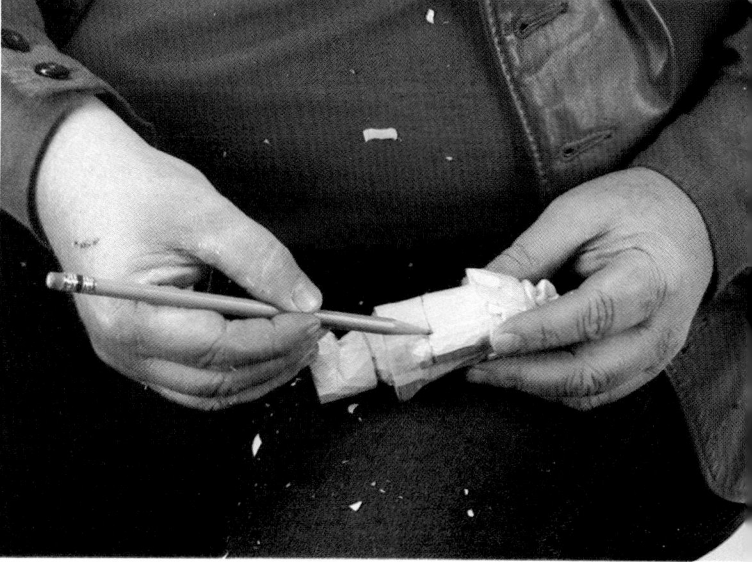

We will draw it above the sleeve line, using our fingers as a depth gauge.

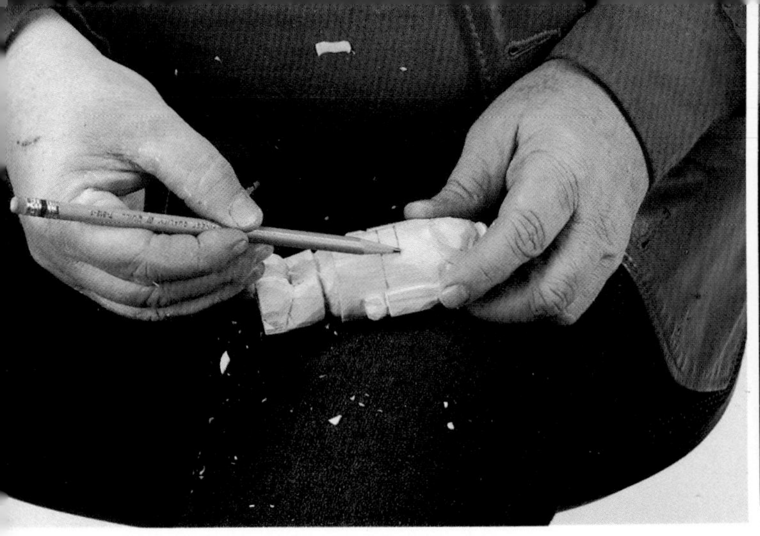

Do the same with the top of the belt.

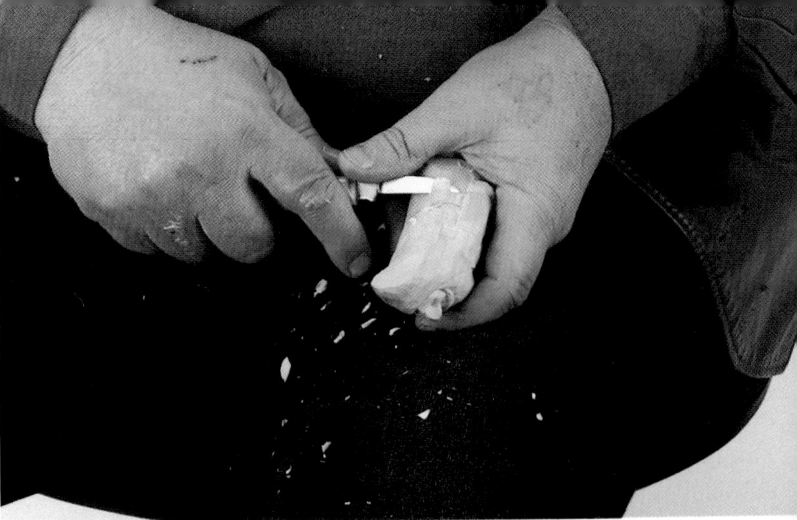

Carve the coat back into the stops.

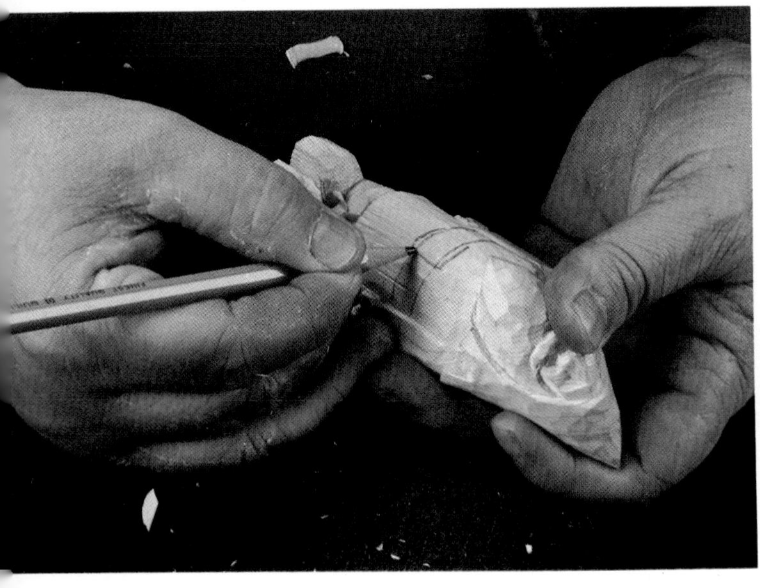

Draw a buckle in the middle.

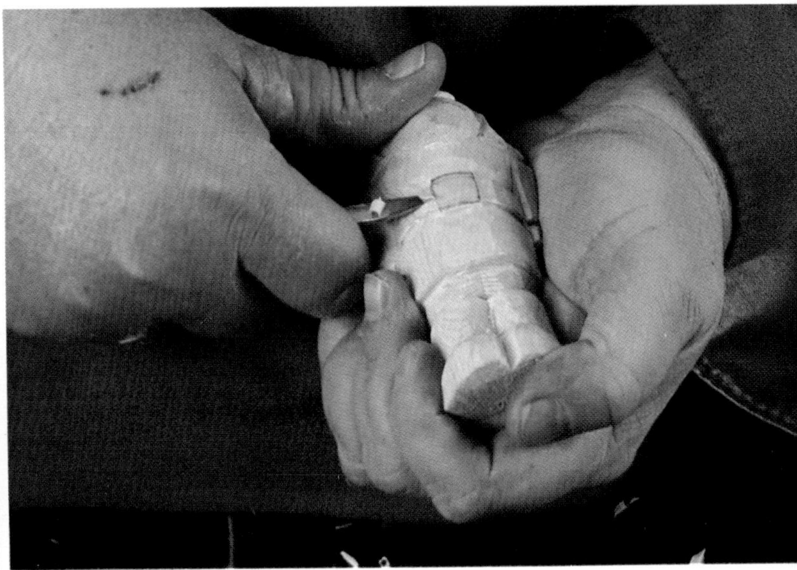

Carve the detail into the buckle

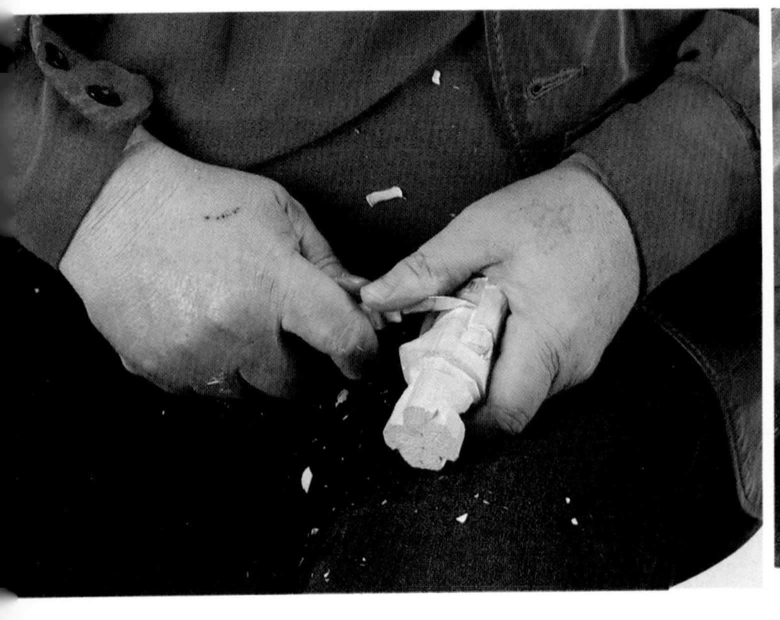

Carve a stop into the top and bottom edges of the belt.

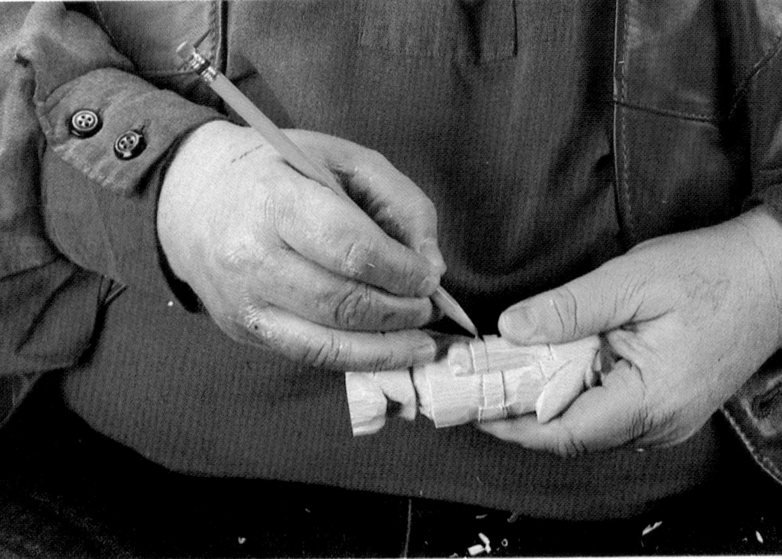

Draw in the trim on the sleeves. Be careful not to meet previous lines.

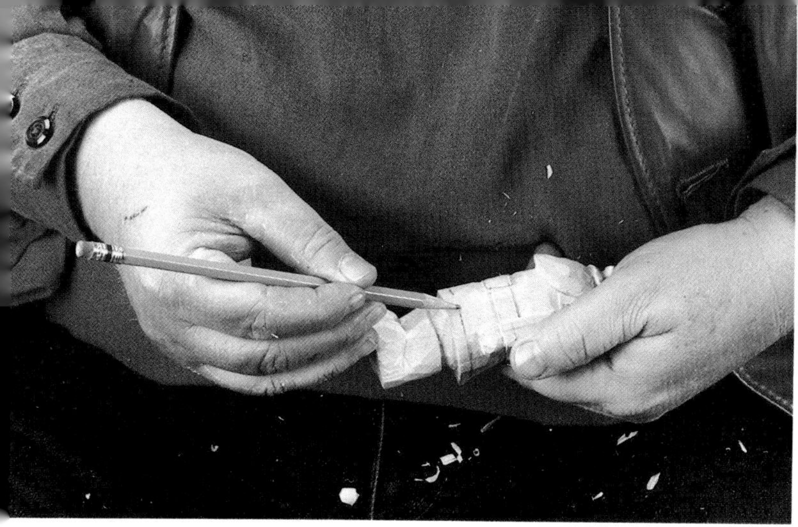

Draw another trim around the bottom of the coat...

Clean up the carving. Go over it until it pleases you.

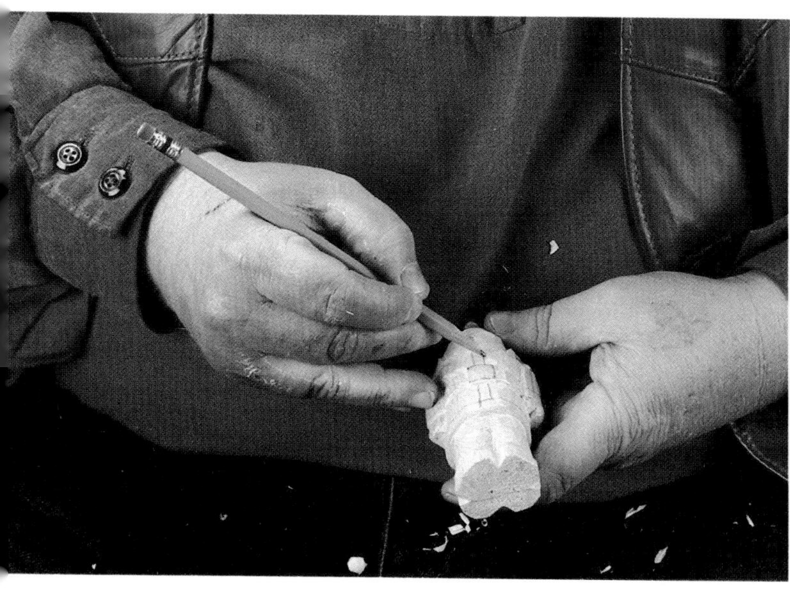

and up the middle of the front. This gives it the traditional Santa look.

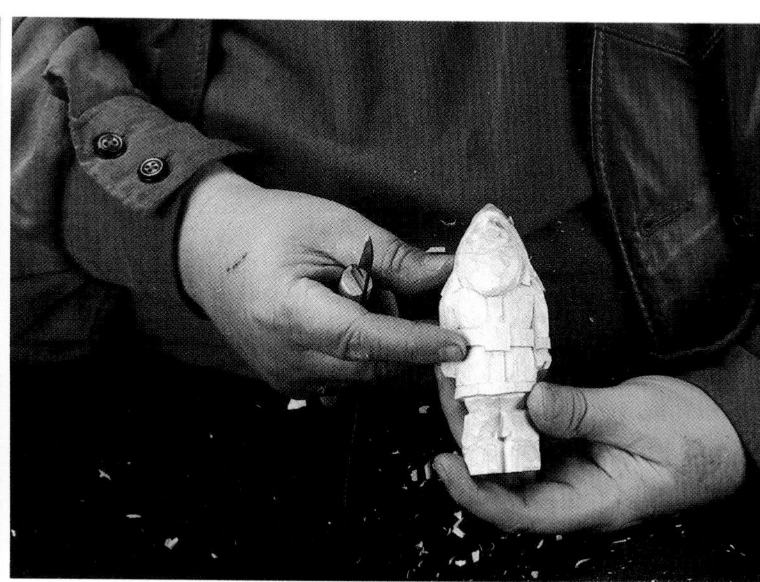

As I was doing this I decided on a rounder, friendlier beard.

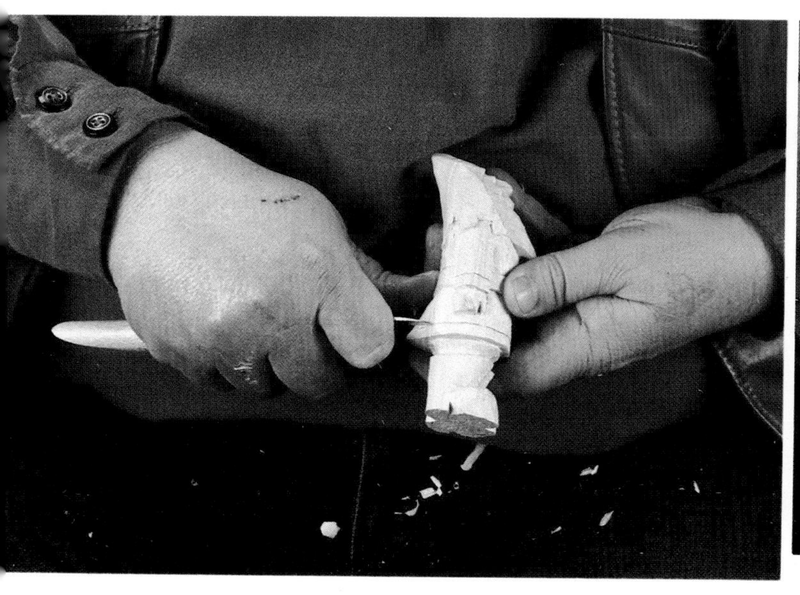

Cut a stop into the lines you have drawn.

I also thought Santa would seem friendlier if the moustache were divided in the middle.

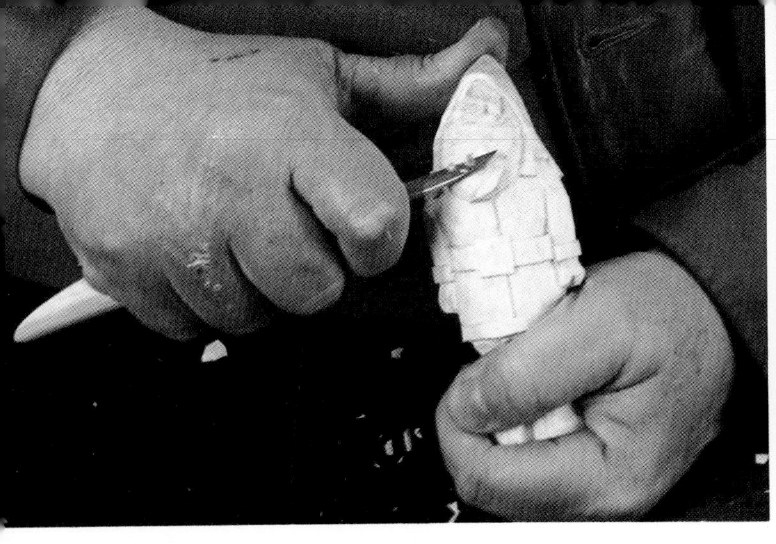

Groove out on either side of the mouth to bring the chin out.

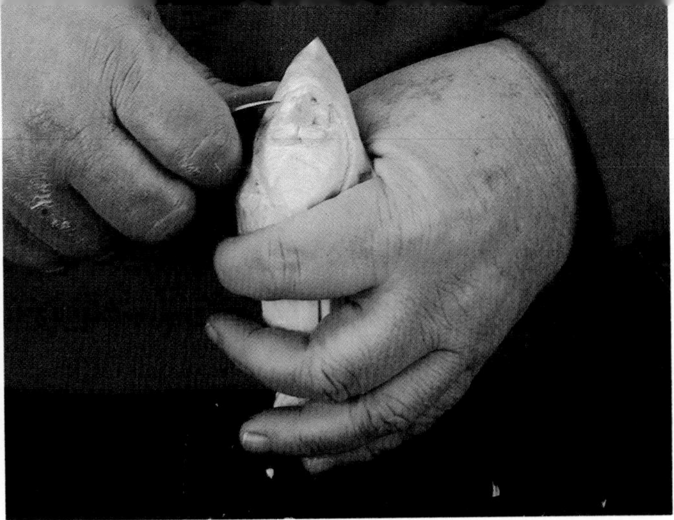

A small slice in the corner of the eye gives it definition.

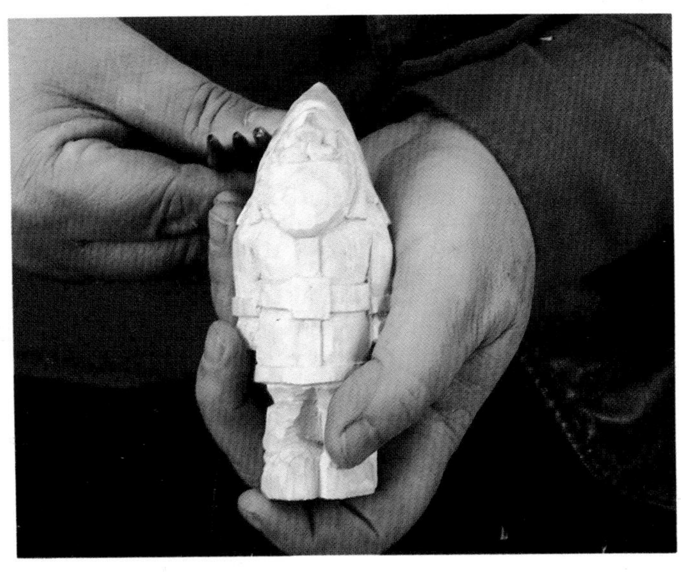

For the eye I use a nailset. They come in different sizes and have different effects. The smaller the nailset the more "beady" the eyes will appear.

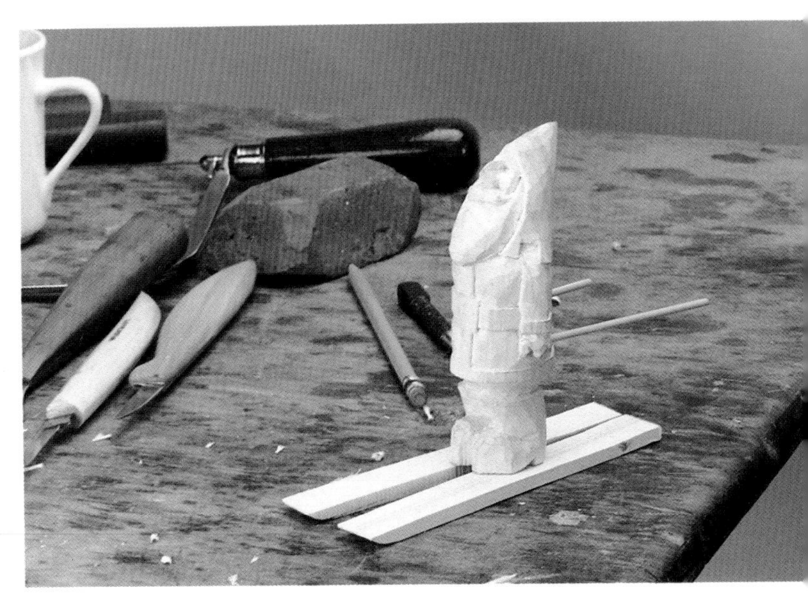

The skis for Santa are from a thin slat of wood split down the middle and slightly upturned in the front. The ski poles are thin dowels inserted into holes drilled into the back of Santa's mittens.

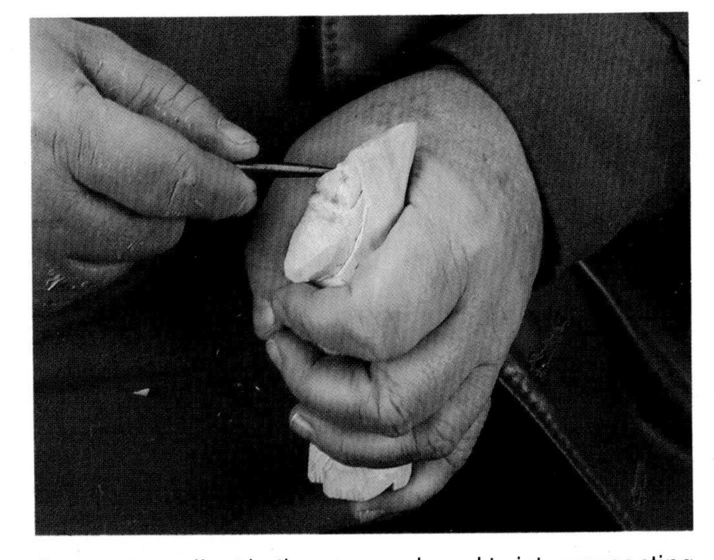

Center the nailset in the eye, push and twist, compacting the wood.

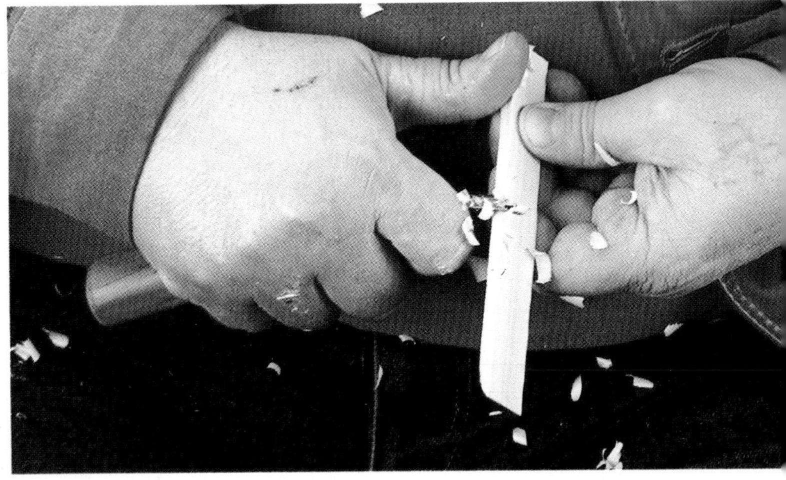

The skis are trimmed down with a knife prior to painting because carving marks take the paint better than saw marks.

Painting Santa Claus

The paint I use is alkyd thinned with turpentine. This gives good penetration without much bleeding. The brushes are a good grade, mostly sable or camel. Brushes are like cars. Usually the more expensive they are the less trouble you have.

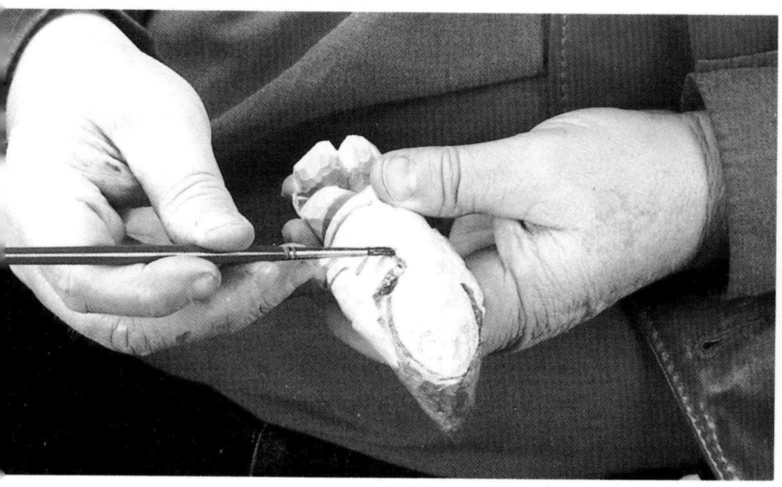

When the carving is finished, the Santa is ready to paint. Begin with the red paint and start painting where the red will meet another color. I am using vermillion red here, but any good red will do. Start on the hood around the face...

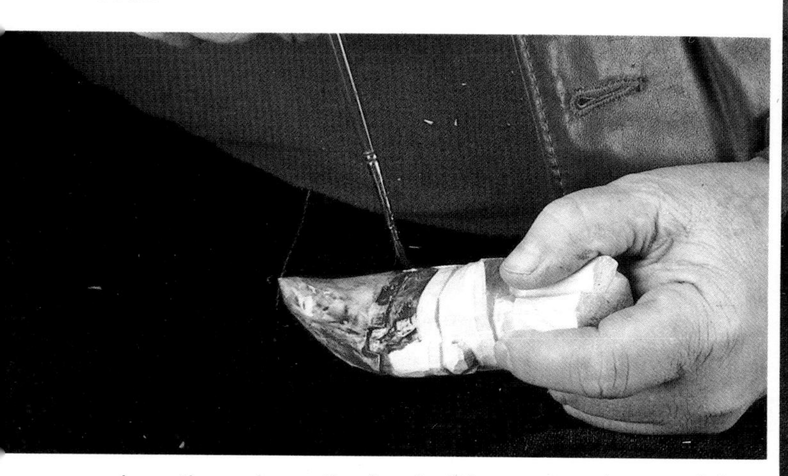

and continue down the front of the coat and around the belt. By painting these "border" areas, the likelihood of colors bleeding into each other is reduced.

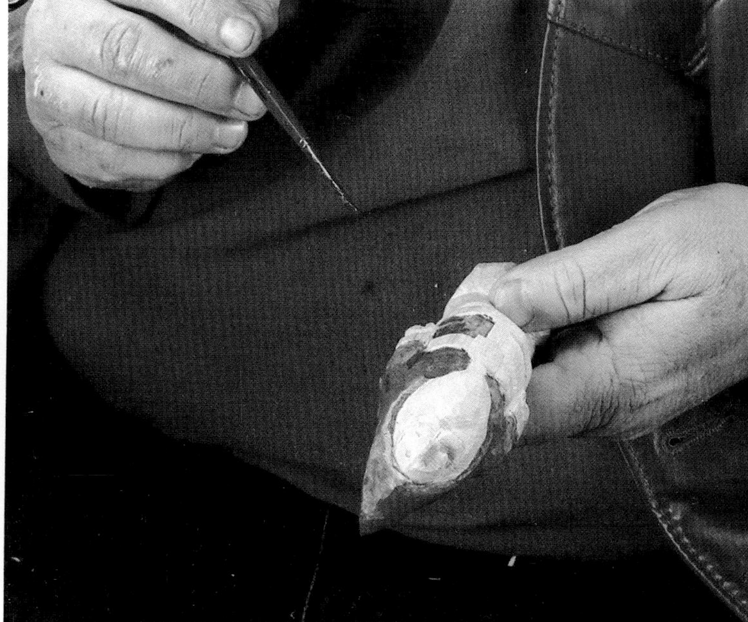

When you've bordered the top part of the coat, fill in the spaces with red.

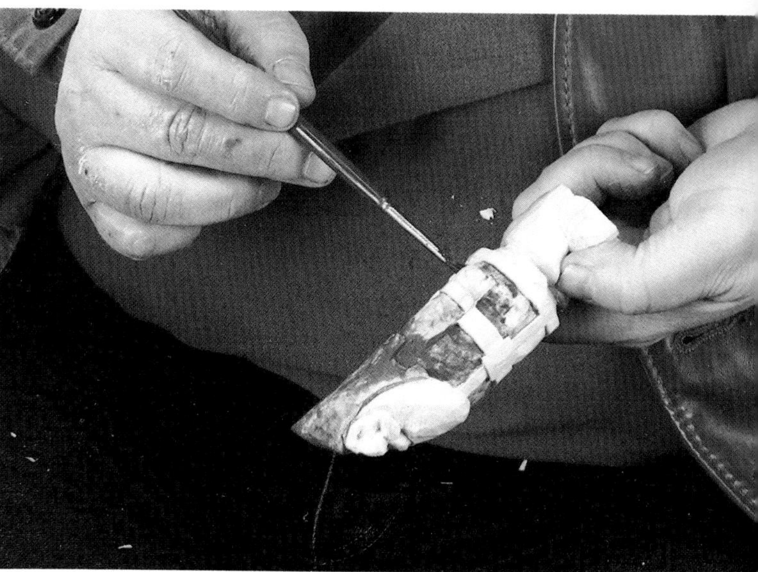

Continue with the red below the belt.

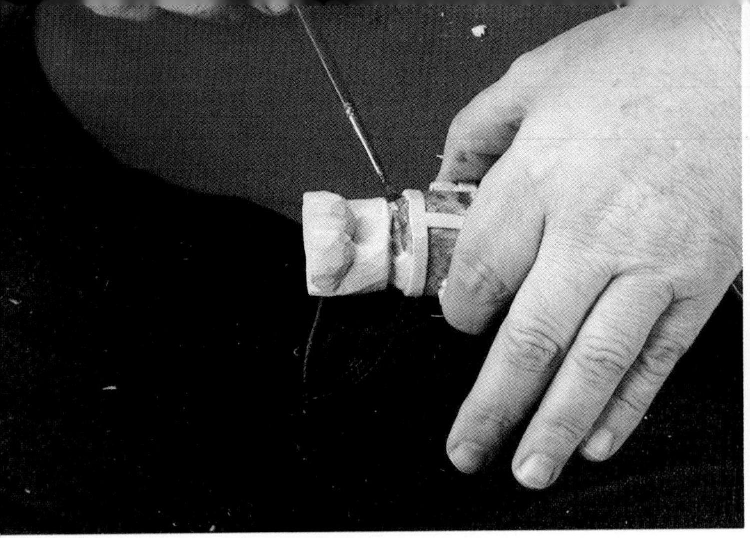

Next, apply red paint to the britches above the boots.

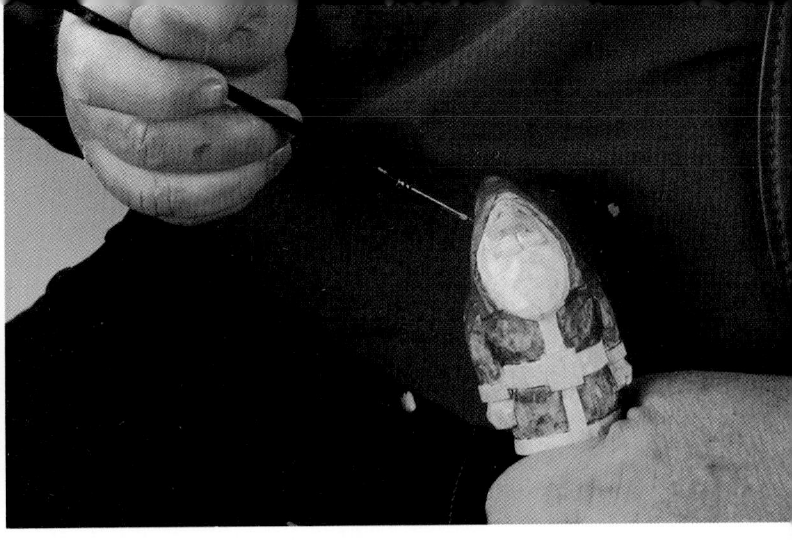

For the flesh tones I take a commercial flesh tone paint and add a little raw sienna. This takes away the strong pink tone that is characteristic of most premixed flesh colors. Just cover the face.

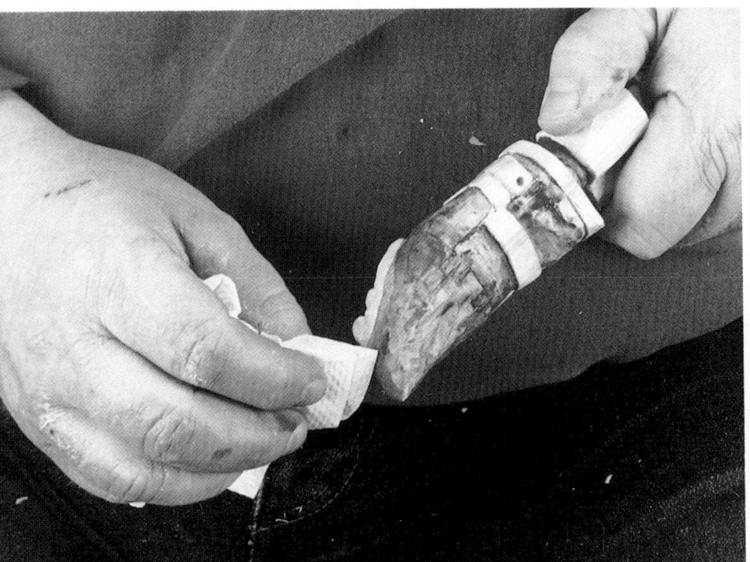

Blot up buildups of pigment like here on the hood...

Red added to the still-wet flesh tone will give Santa the rosy cheeks and cherry nose for which he is famous.

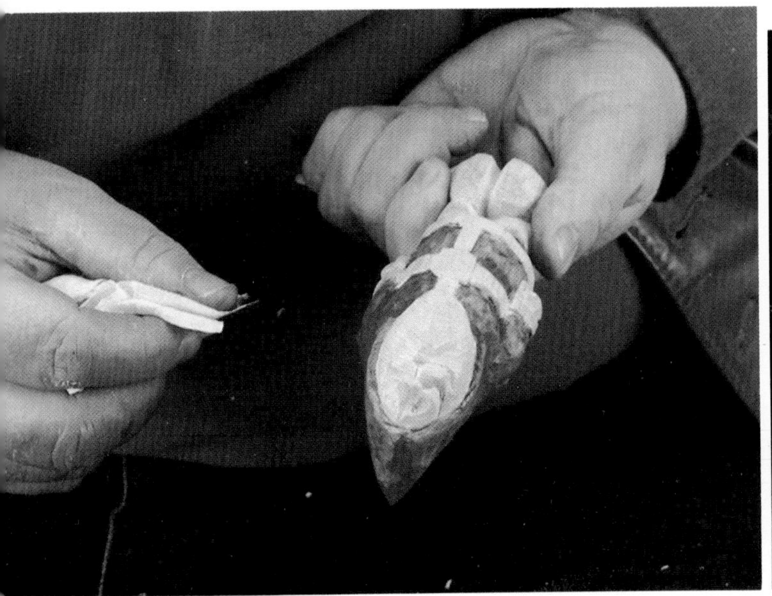

using a paper towel.

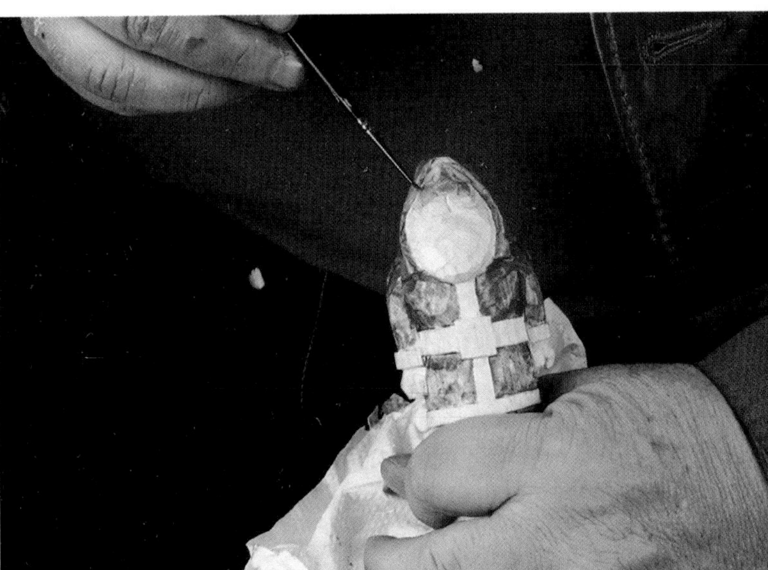

Blend colors together with a thoroughly clean brush.

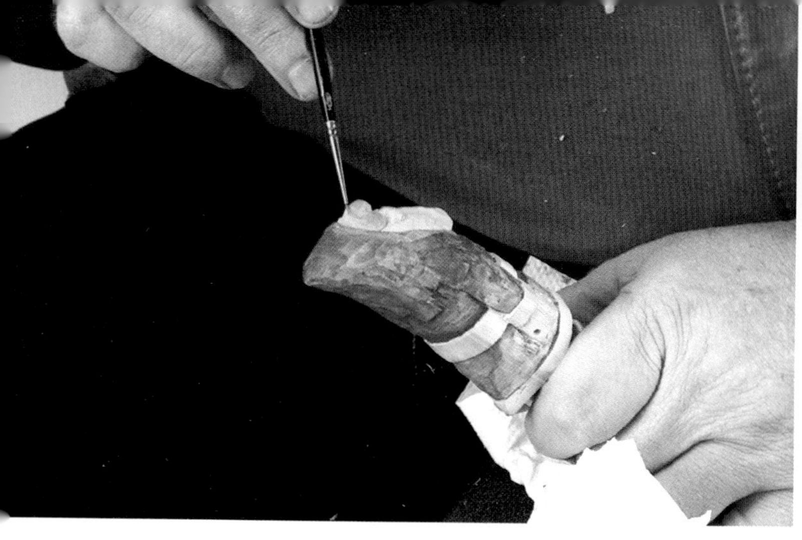

Finally, with a dry brush blend it some more, wiping any pigment you pick up on a paper towel.

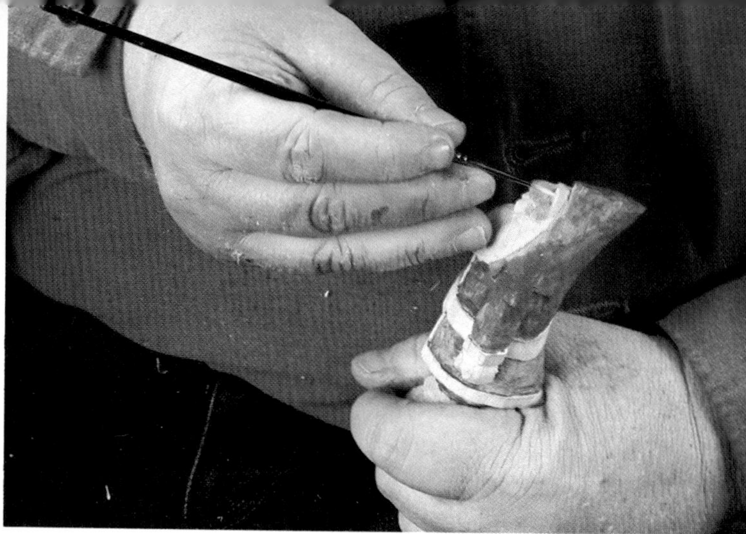

Using that deposit like an artist's palette, it is possible to take smaller amounts and apply them to more delicate areas like the eyebrows.

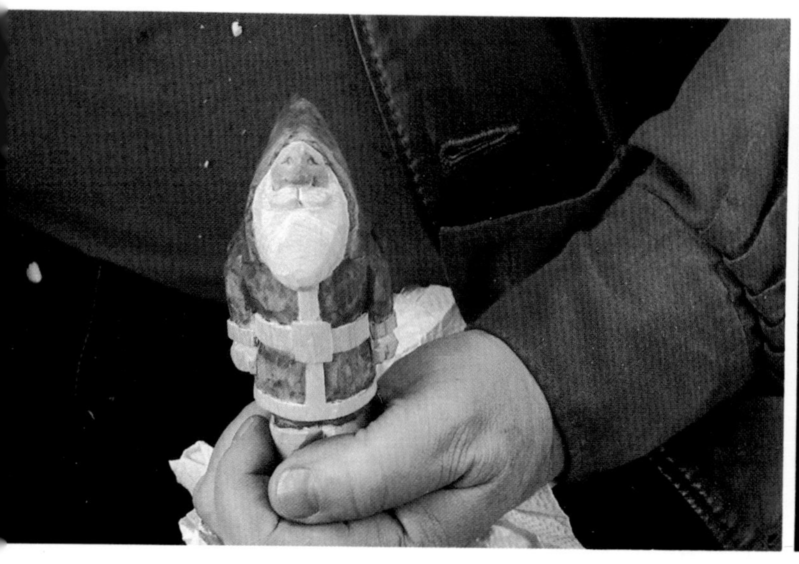

The result is the ruddy, outdoors look typical of someone from the North Pole.

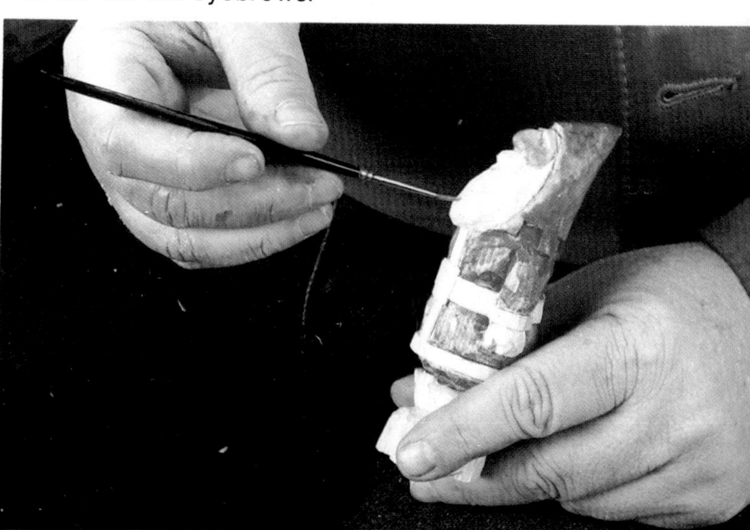

Finish painting the beard and moustache.

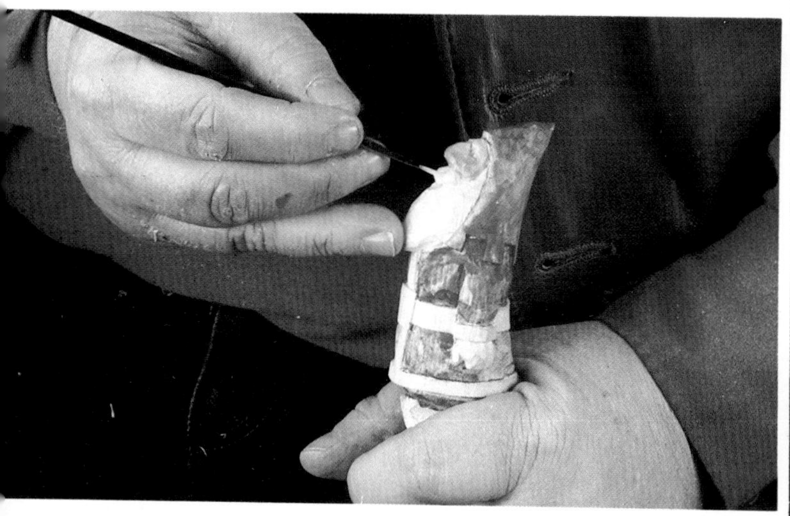

Take some white paint from the cap, where there is a more concentrated pigment. This concentrated white is applied to a large area of the character, in this case the moustache/beard.

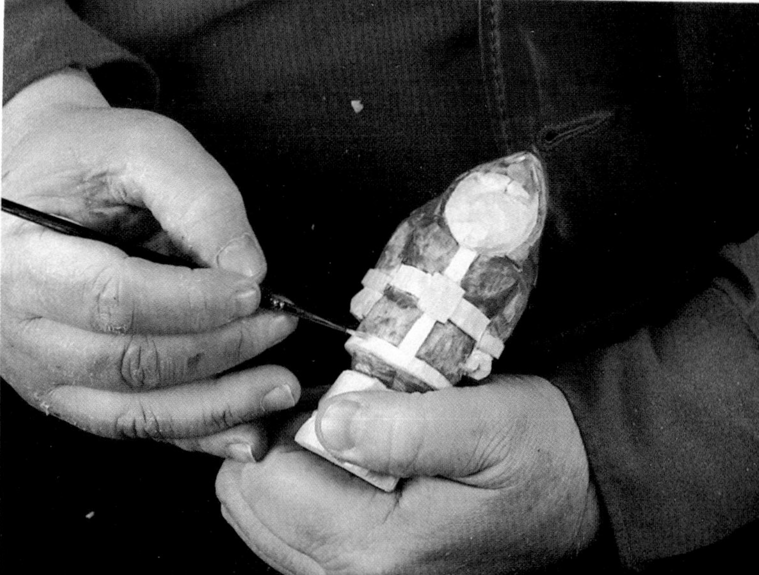

Continue the white around the trim of the coat.

Paint the belt and boots black, leaving the buckle unpainted for the moment.

the belt buckle yellow...

Wipe off the excess black with a paper towel.

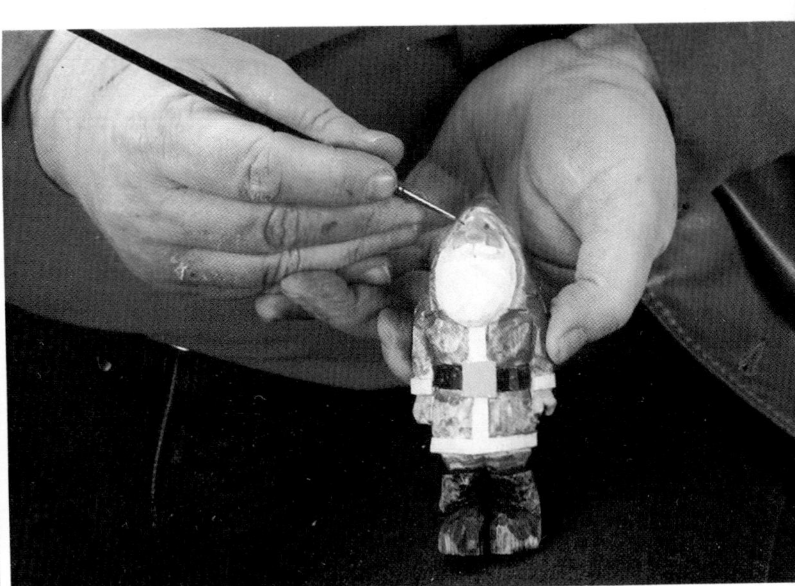

and the eyes blue. Because the iris on this Santa is so small we have chosen a bright blue...

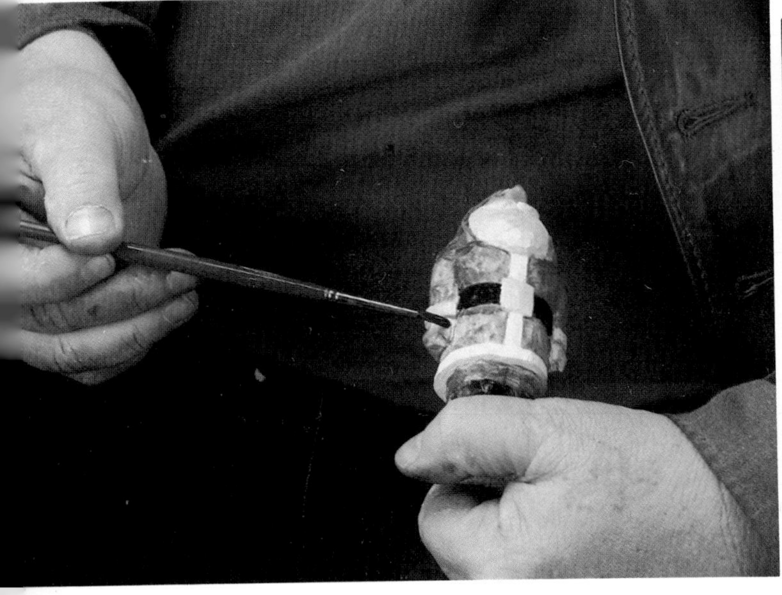

Paint the mittens green...

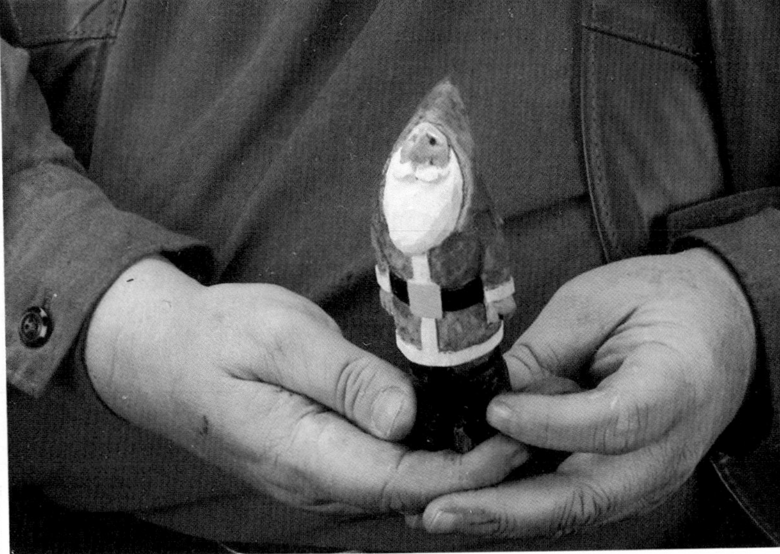

so it will show up.

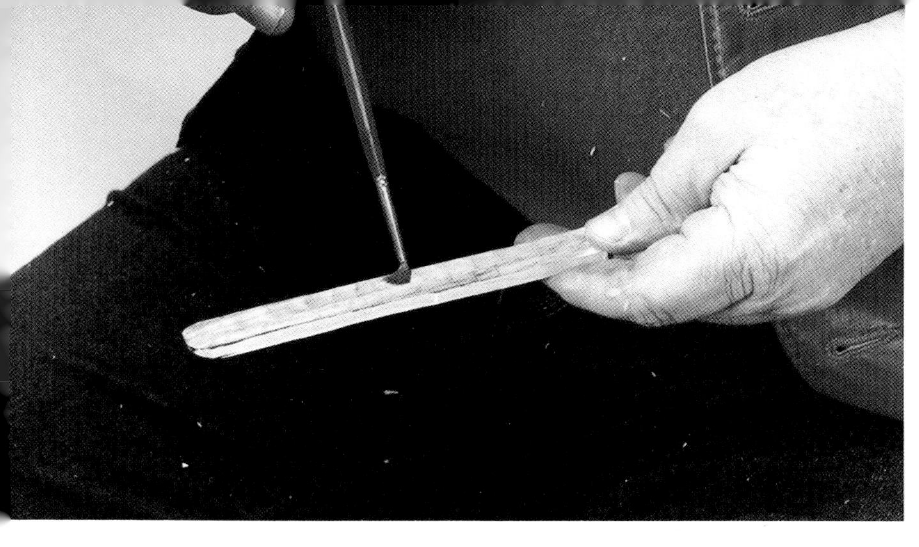

The skis and the poles are painted with a diluted burnt sienna, so it has the effect of a stain.

Four views of our finished Santa.

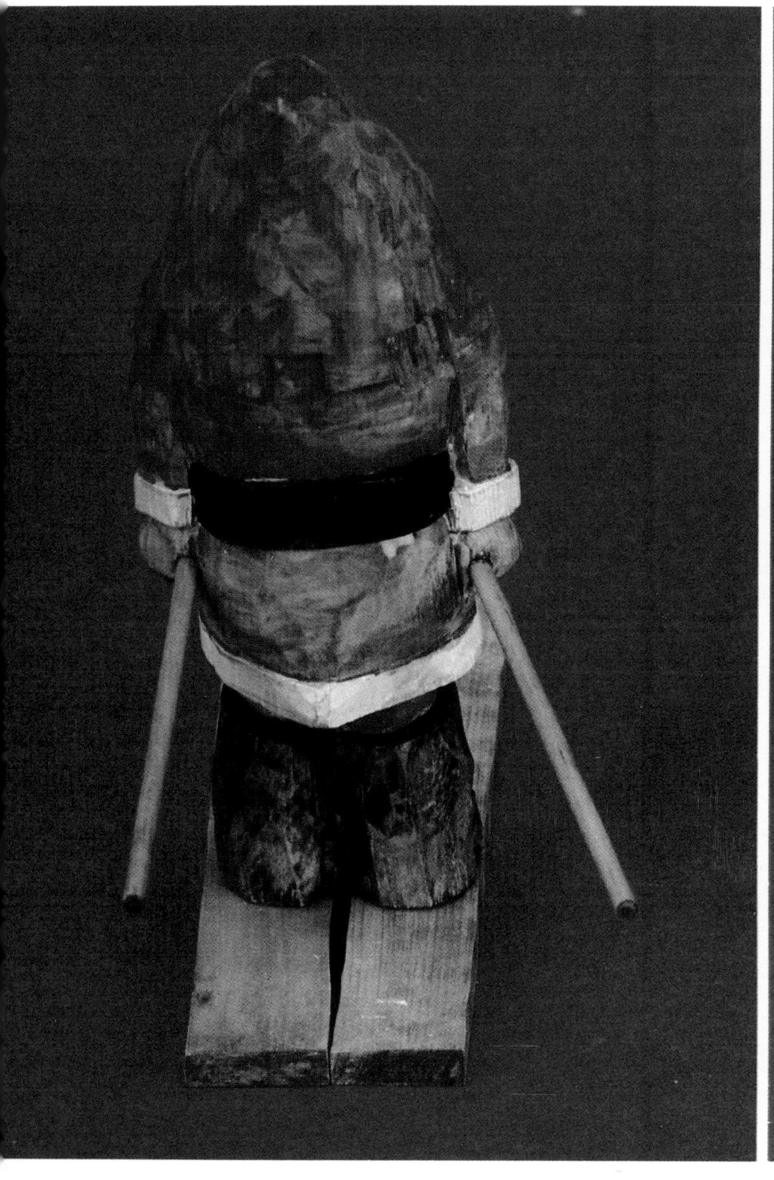

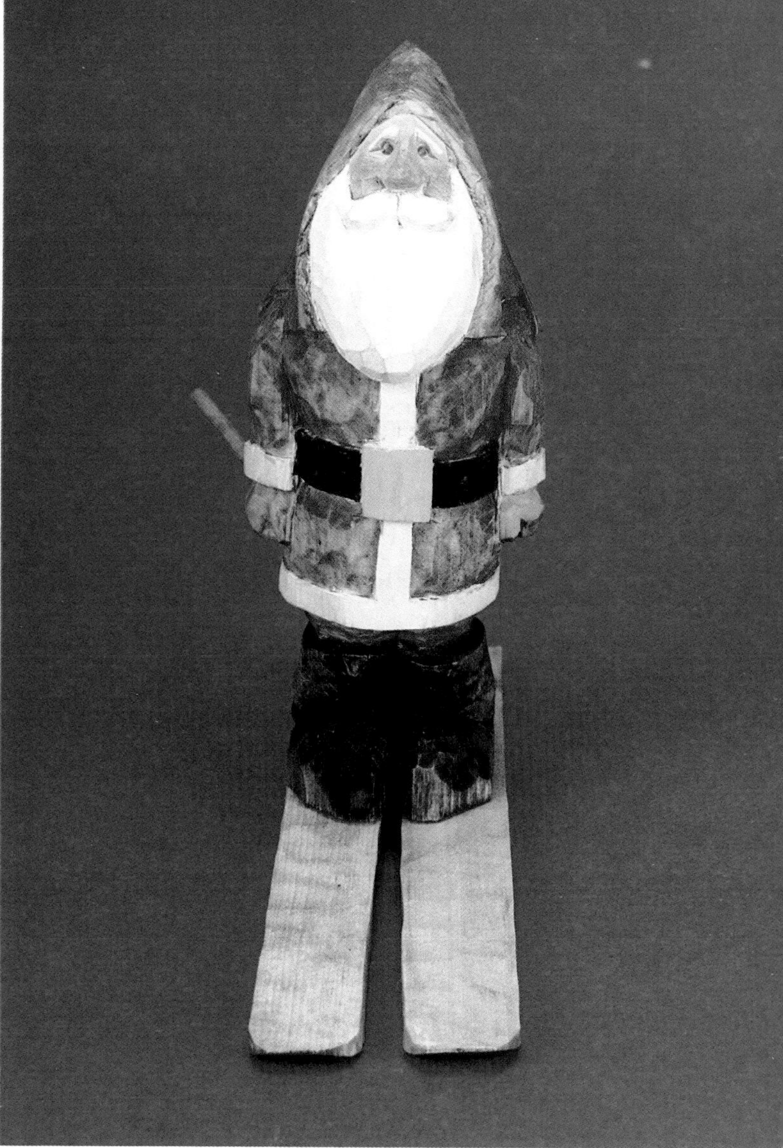

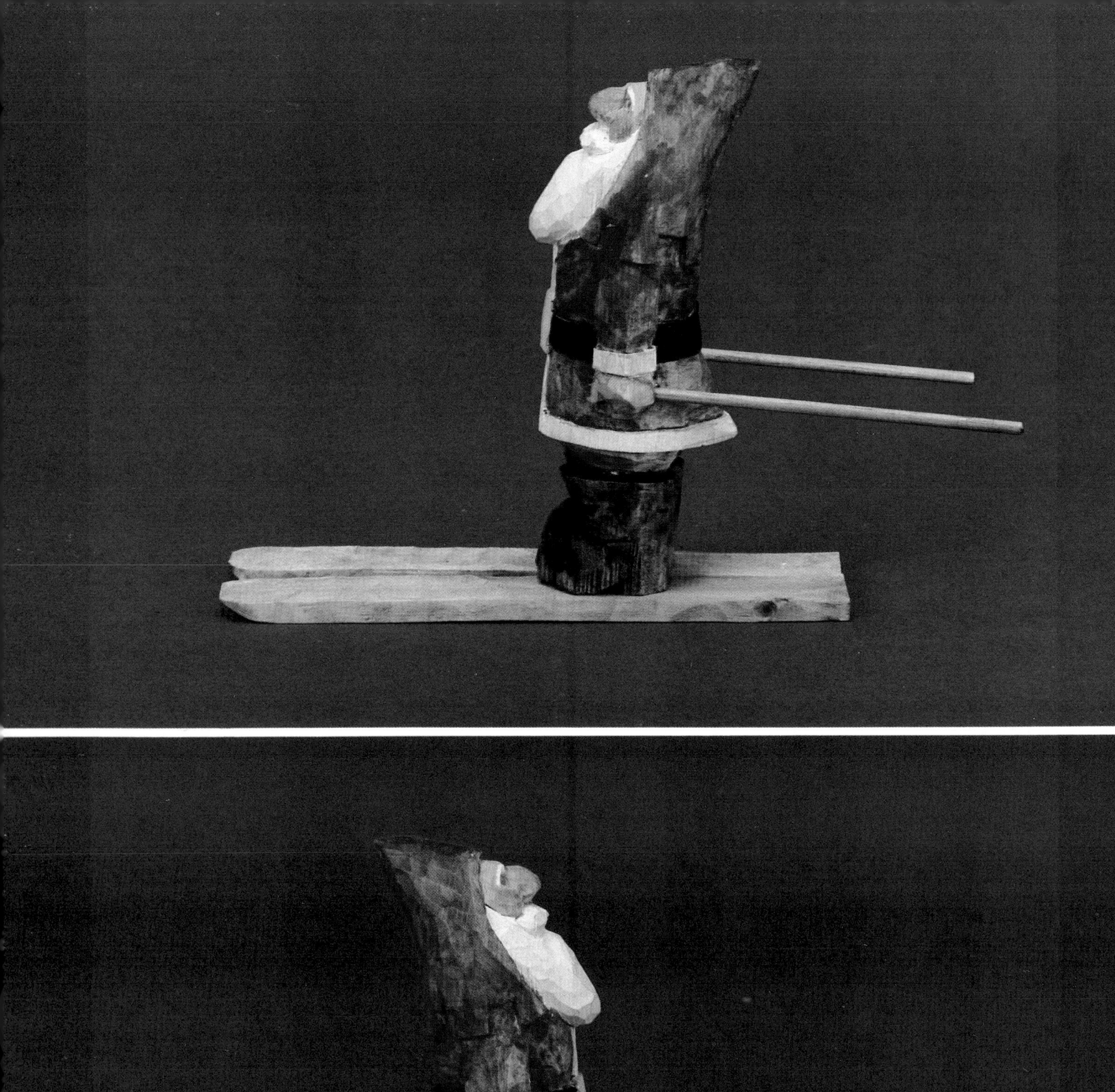

Views of other Santa figures

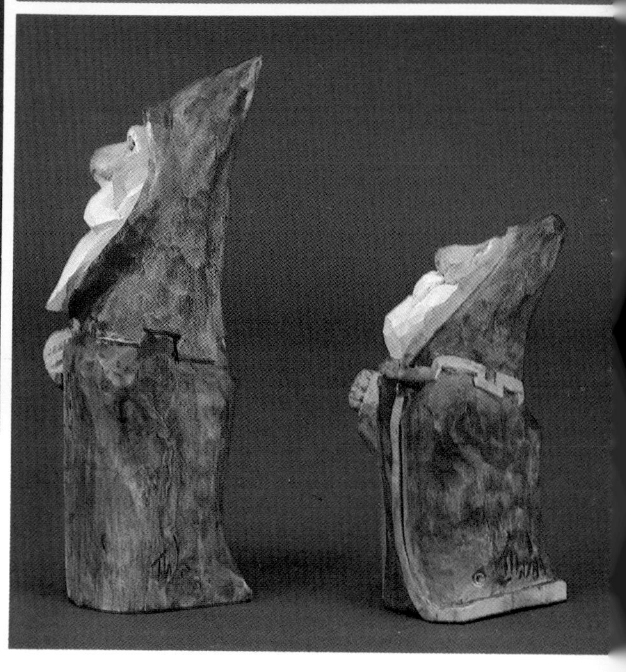

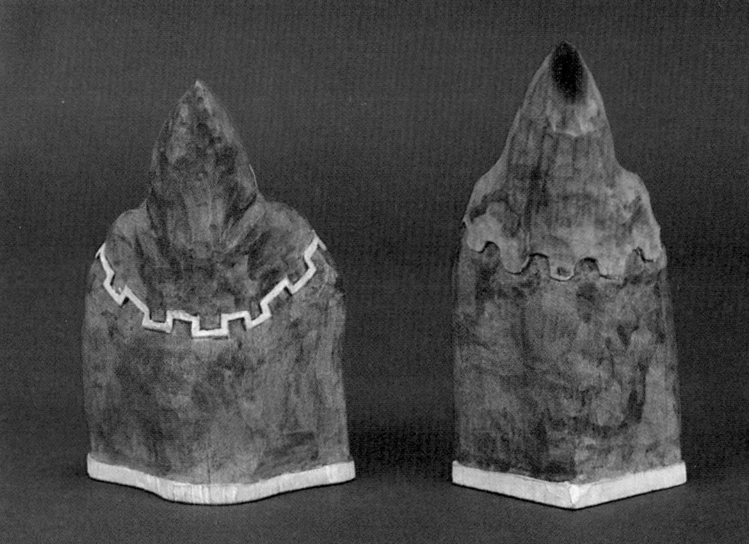

Carving a Robed Gnome

Using the same general method as in the Santa carving, many figures can be carved. The only limitations are your imagination and your time. At the back of this book you will find many examples.

Here we will continue with a gnome. A smaller project than the Santa, it still uses the basic ideas and skills we used there. Again we start with a length of square dimensional lumber, in this case a 3¼ inch length of 2 x 2.

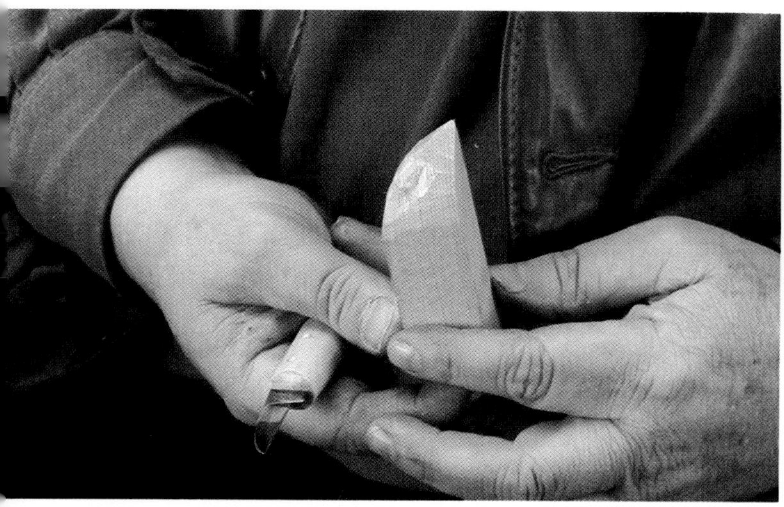

Begin by rouding off toward one corner (as with the Santa carving). Include the general shape of the nose as you do this.

Put in the line of the upper lip, beginning with a stop cut and carving to it. This is an exaggerated, broad feature.

Bring the side of the hood down, cutting a stop and silhouetting the face.

Round up the hood, bringing the sweep of the hood to where the legs would start.

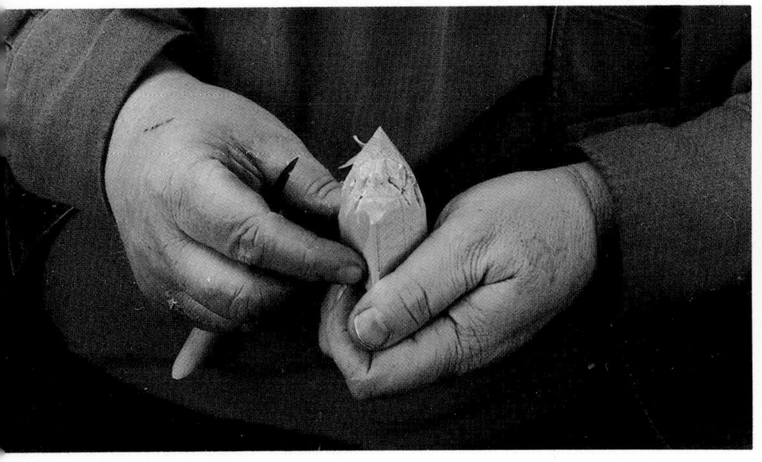

Carve in the beard line.

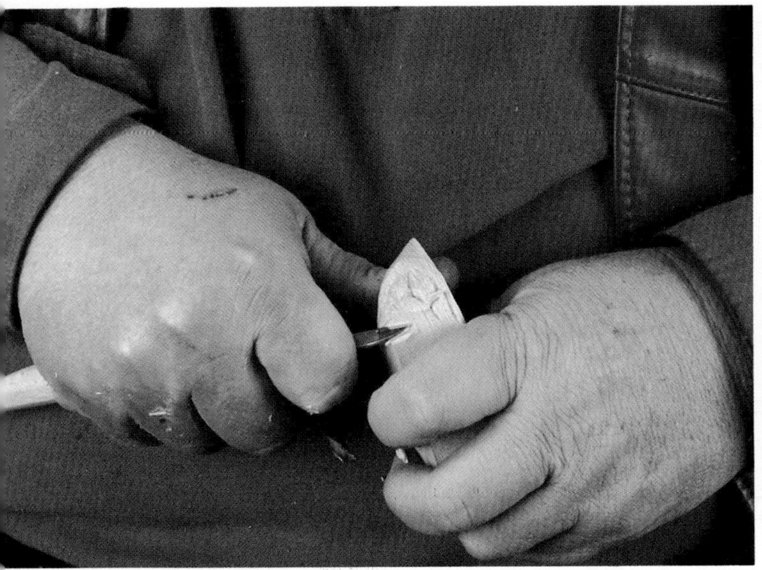

Cut out the bottom lip.

Define the bottom of the beard and cut a wedge.

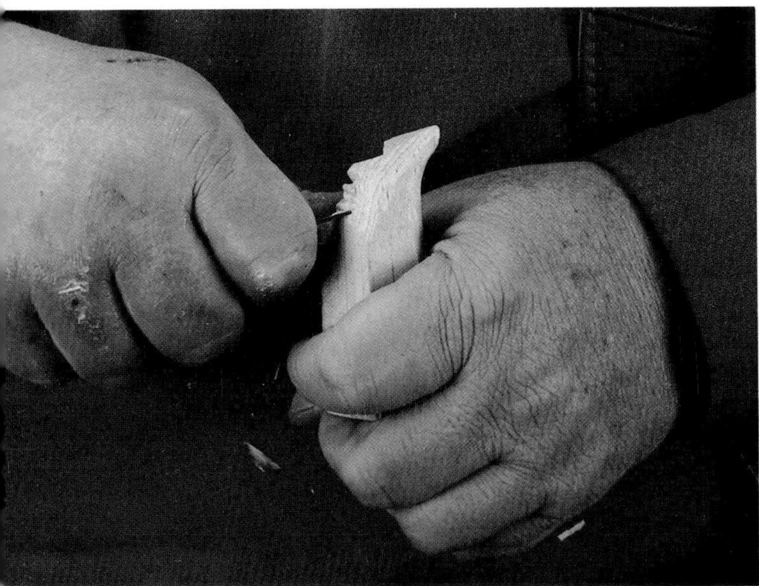

This is the lip in profile.

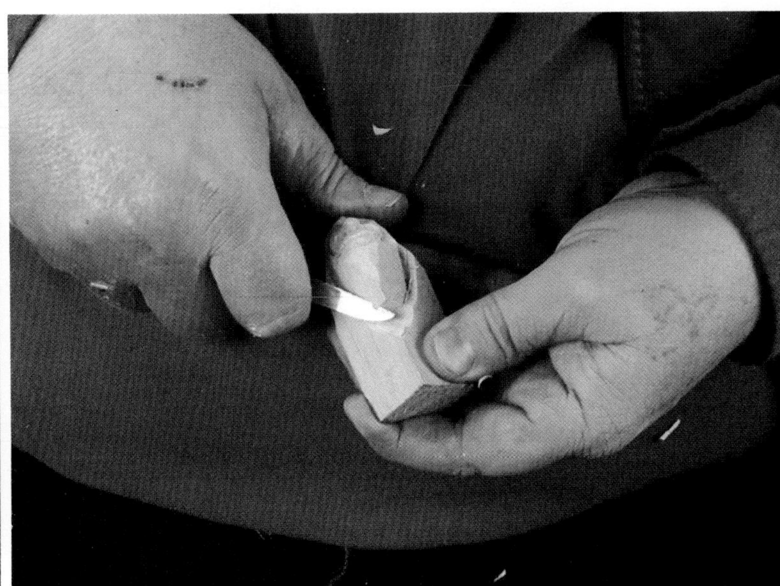

Carry the line around to meet the hood.

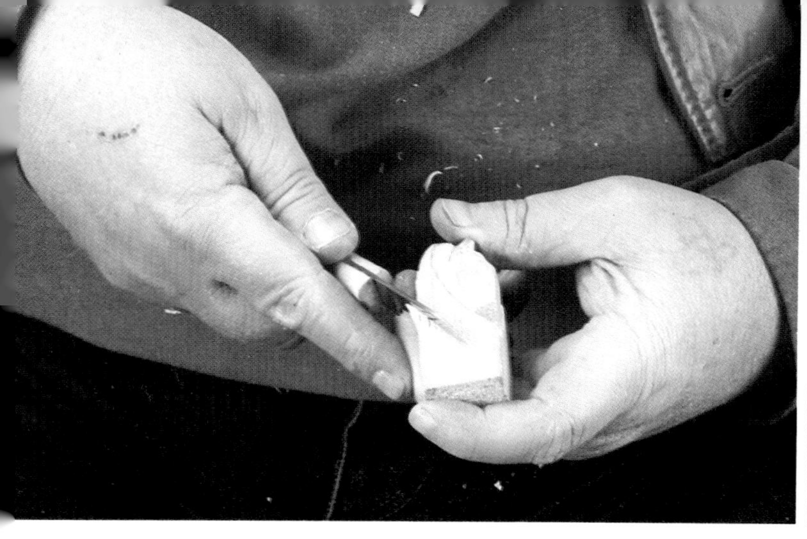

Define the position of the arms. In this model the arms will meet in front of the gnome.

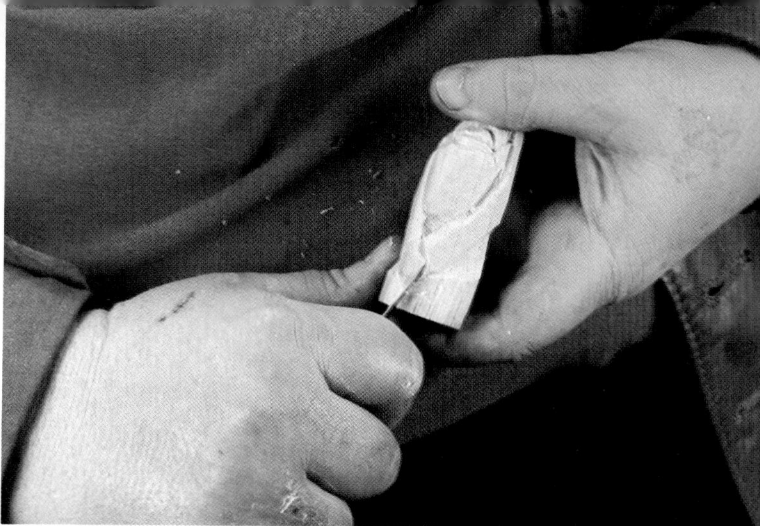

Bring the hands out a bit by silhouetting them.

This same line forms the arms, shoulders, and hands.

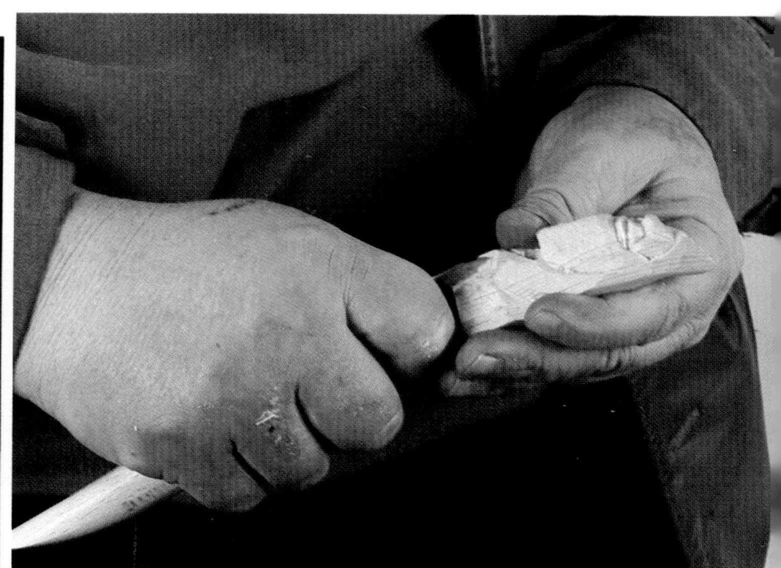

Define the thumb...

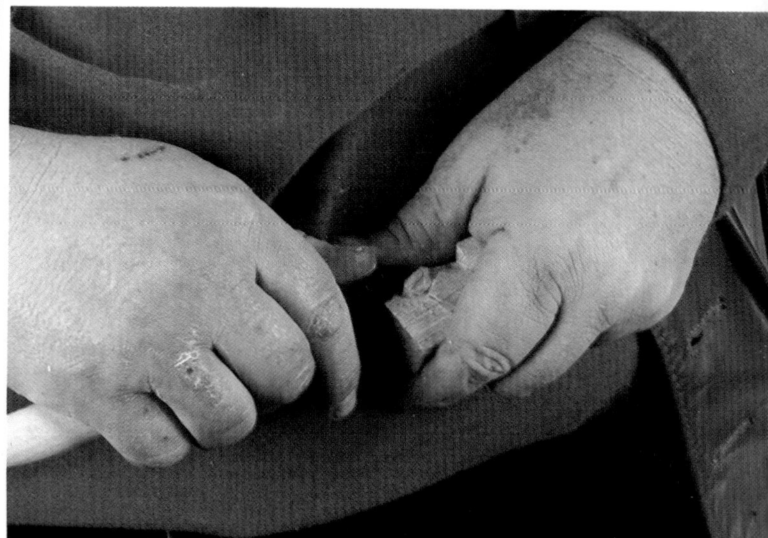

and bring two lines down to form three fingers. Often these characters, like those in cartoons, look better with three fingers than four.

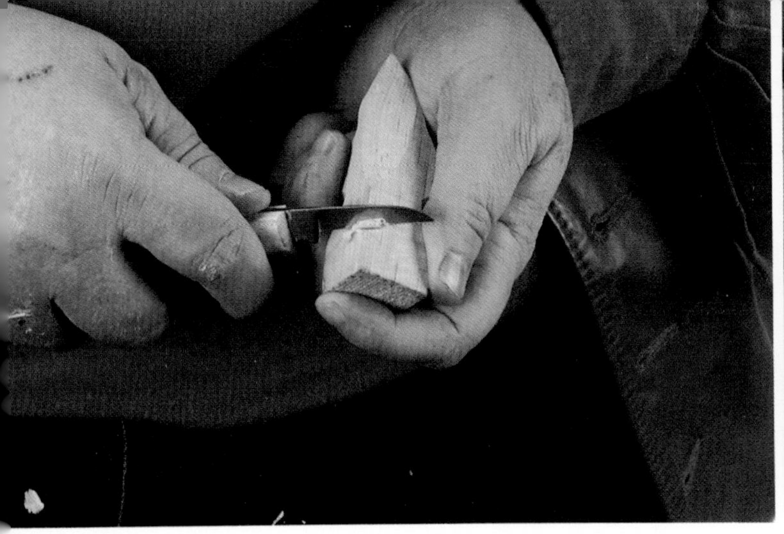

The gnome will be kneeling. A notch in the back...

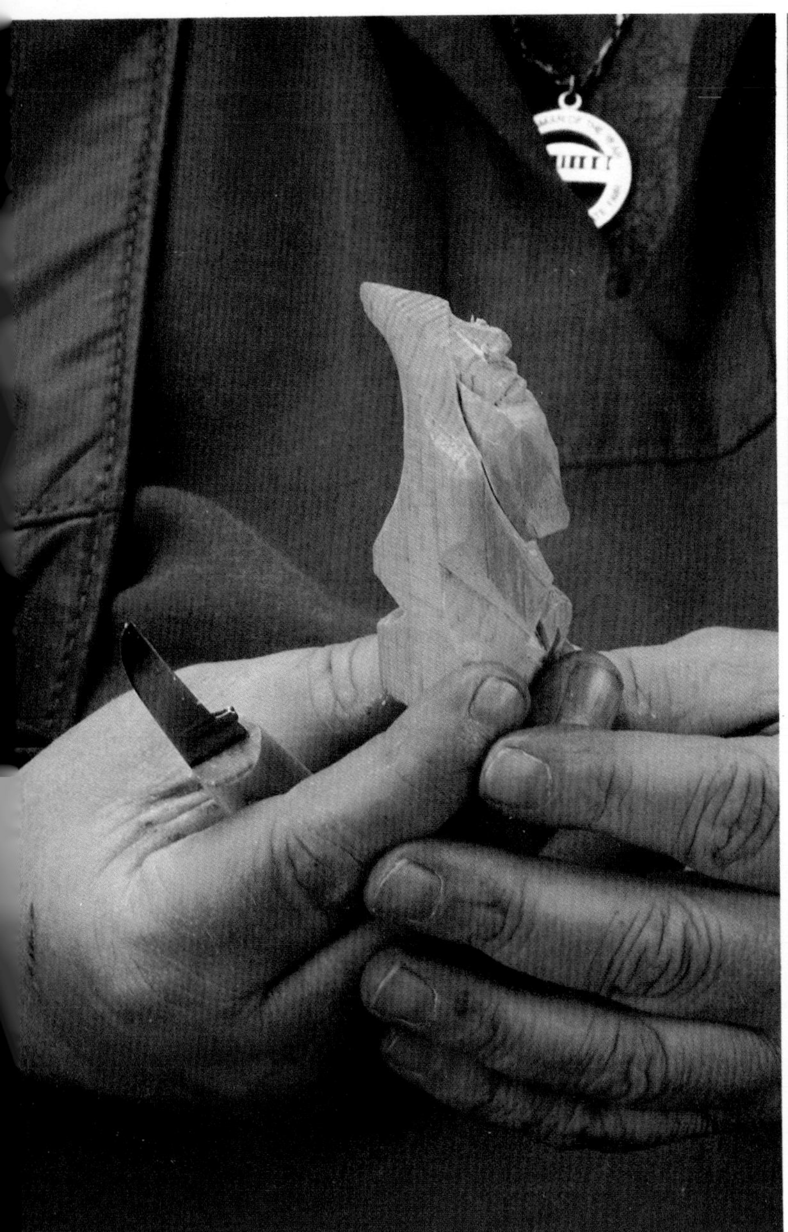

will define that shape.

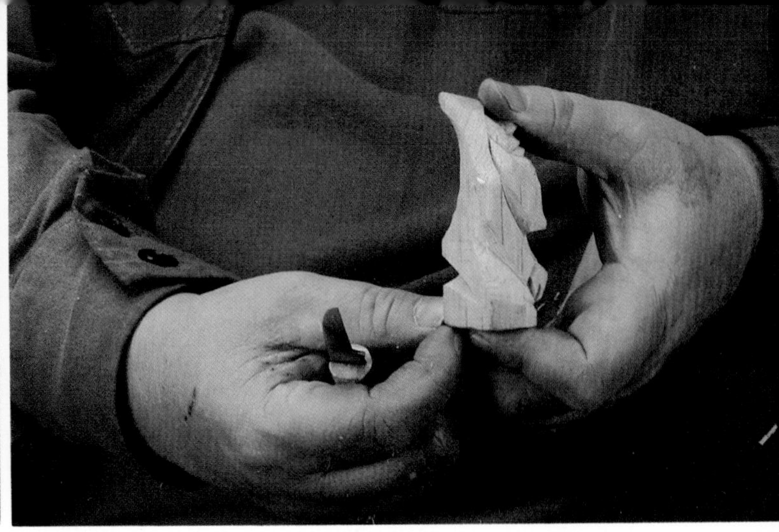

Deepen the notch to shape the inside of the leg.

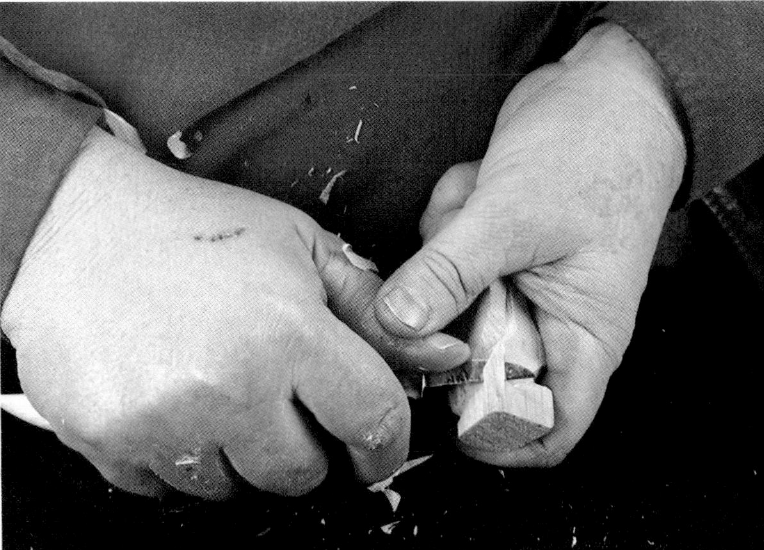

Round the corner and narrow the base.

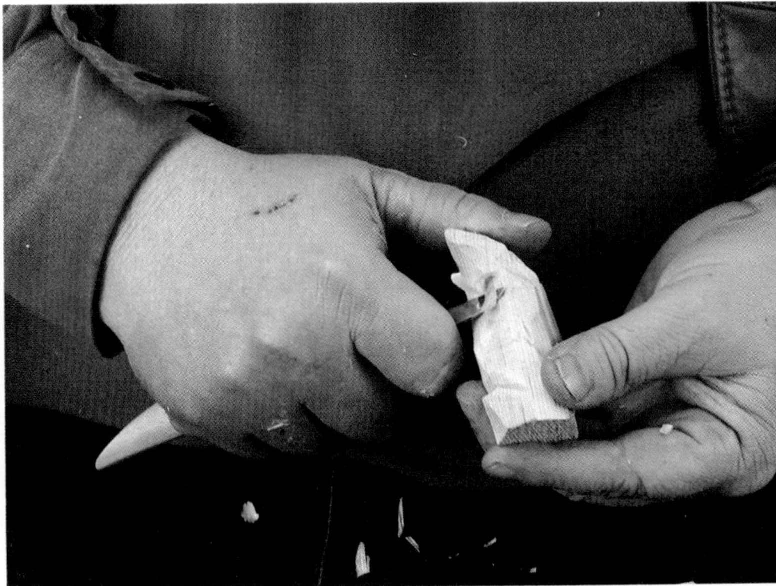

Continue to round and shape.

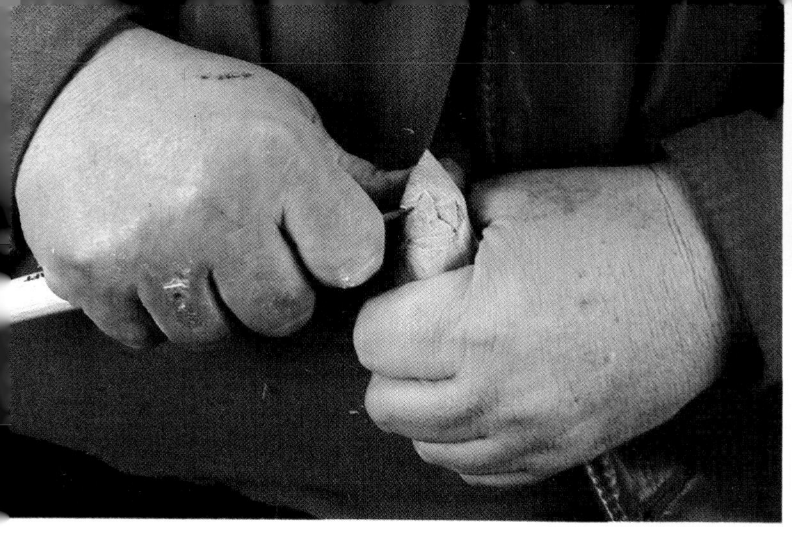

Begin defining the eye by hollowing out the socket.

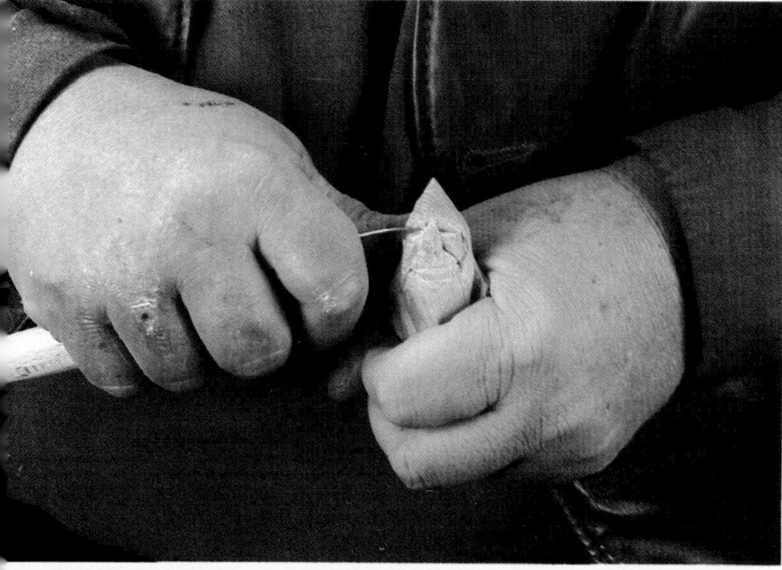

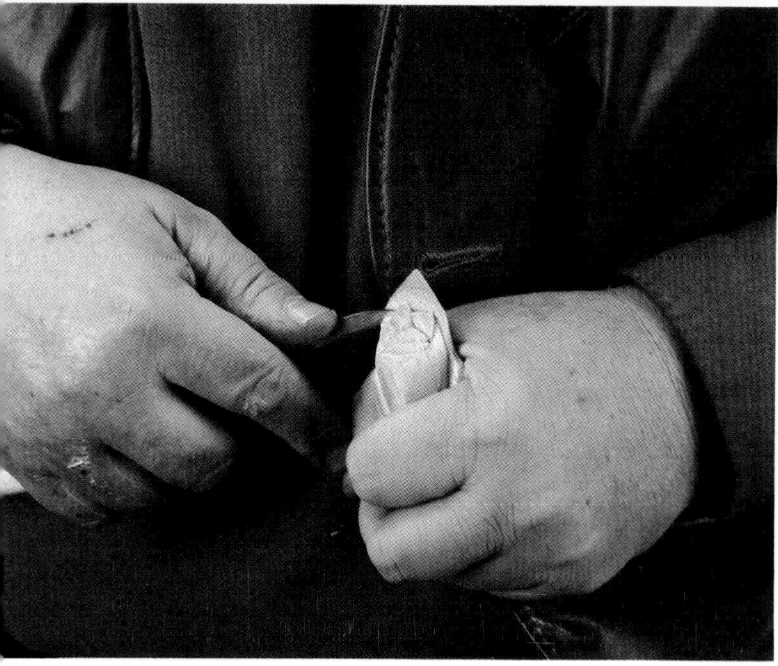

Cut slits for the eyes.

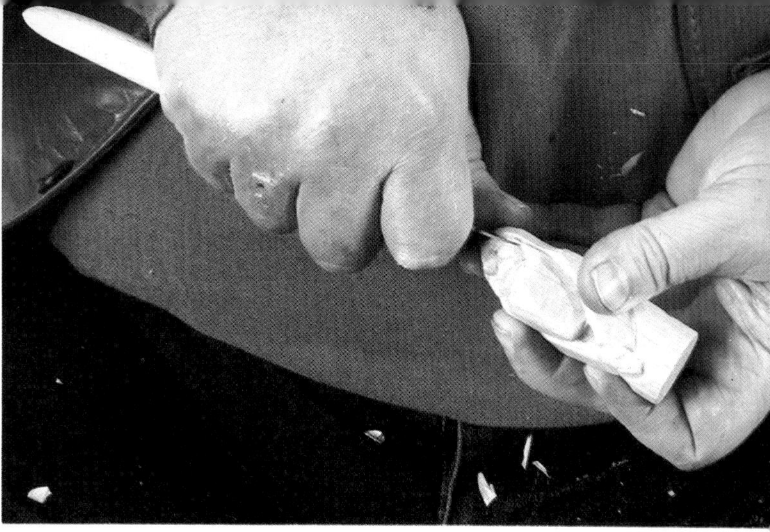

Come in at the temples to give more bulge to the cheeks.

Carve a belt around the back by using a deep v-shaped cut.

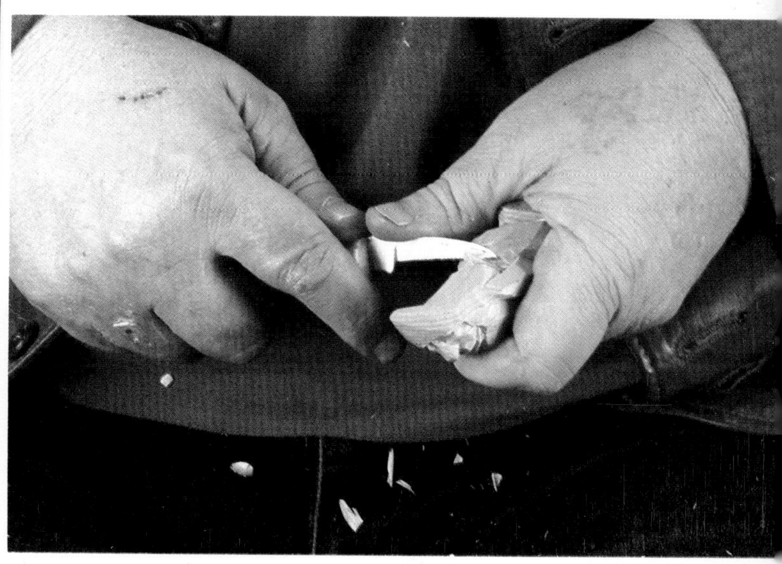

Bring another slanting cut down to the first and leave a rope belt at the bottom of the v-shaped inset.

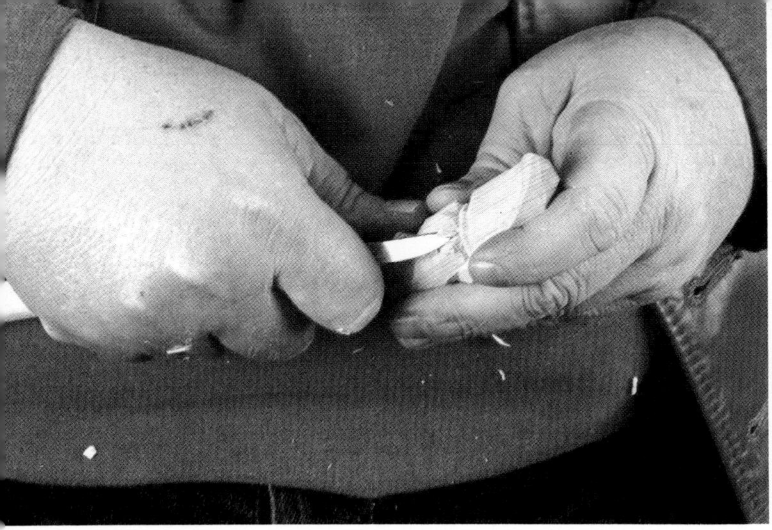

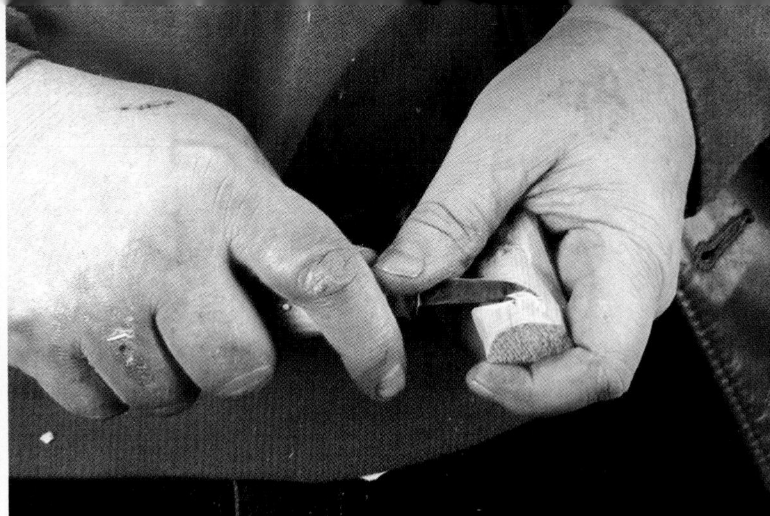

Take some cuts out of the robe around the waist to show the gathering of the cloth.

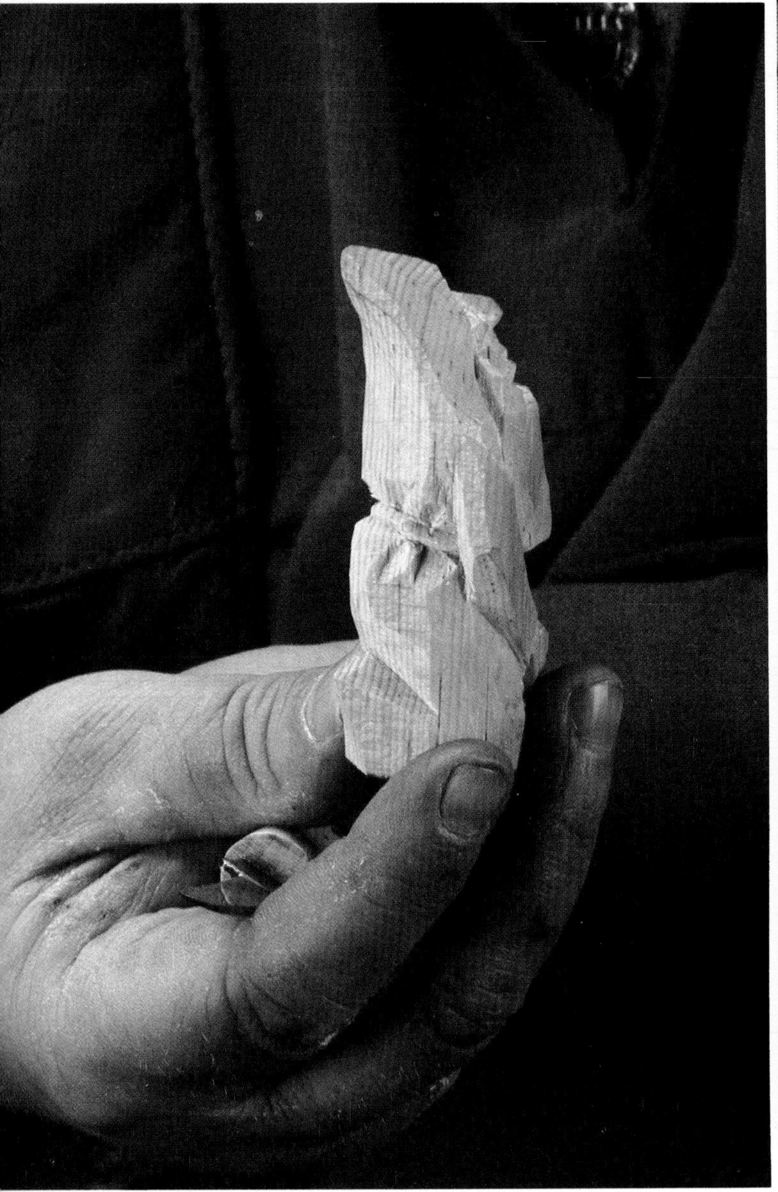

This is how the back will look after the belt is finished.

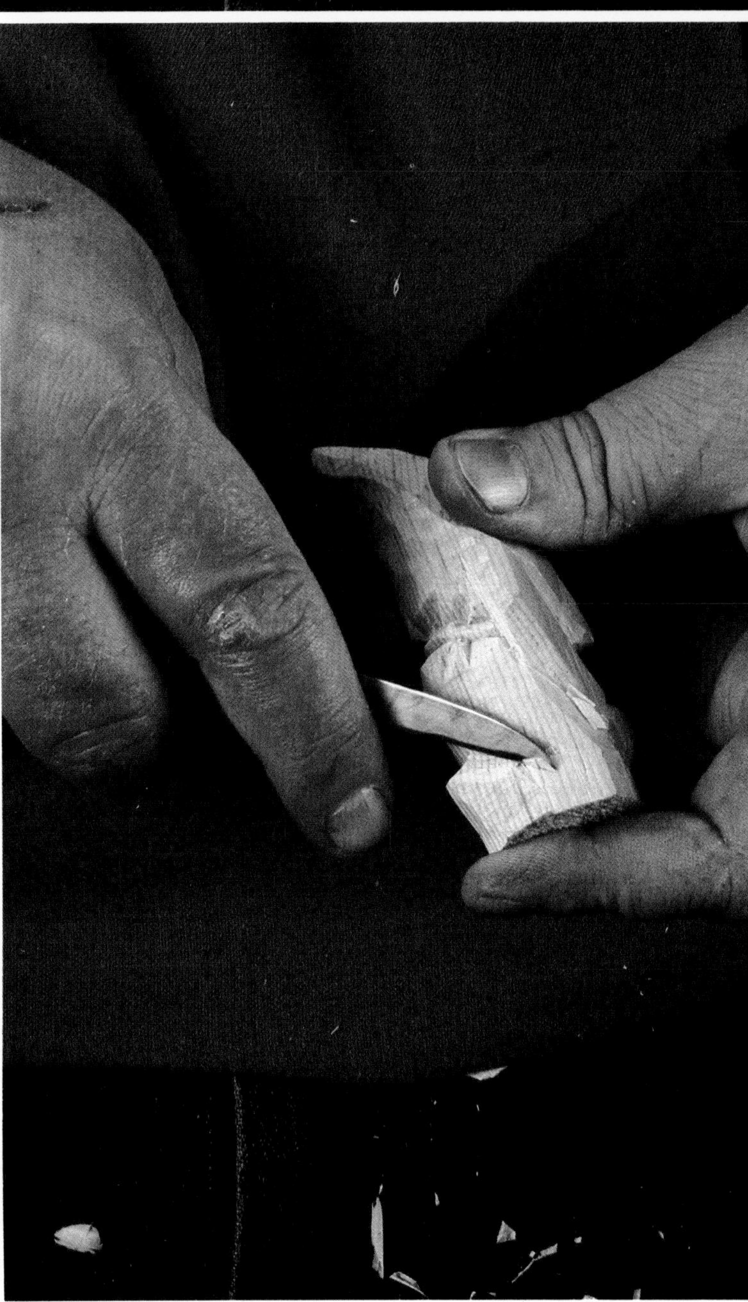

Carve some small slits at the back of the knees to show the fabric folds.

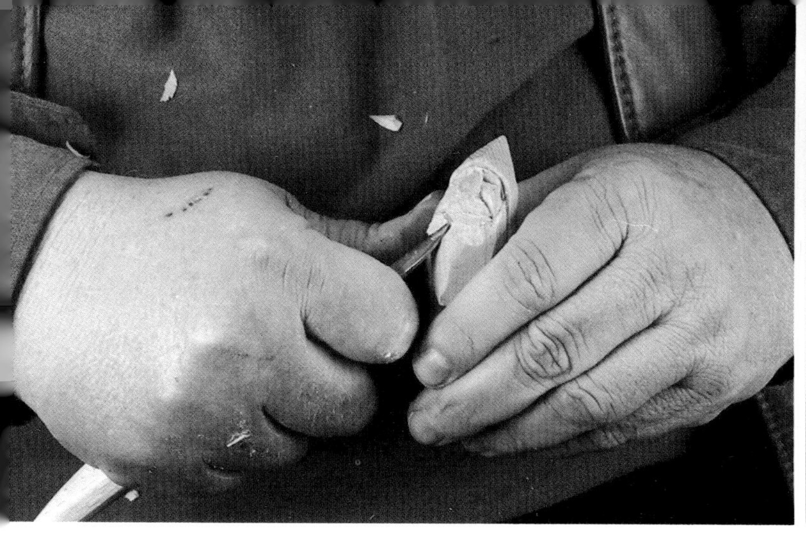

Refine and smooth to give greater definition to the chin.

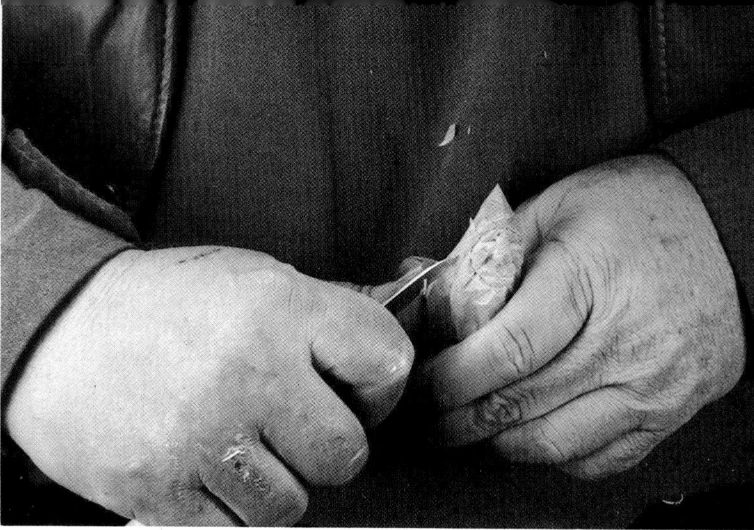

We'll add a border to strengthen the area around the hood. Often if an area doesn't look quite right, it helps to go back and do a little more.

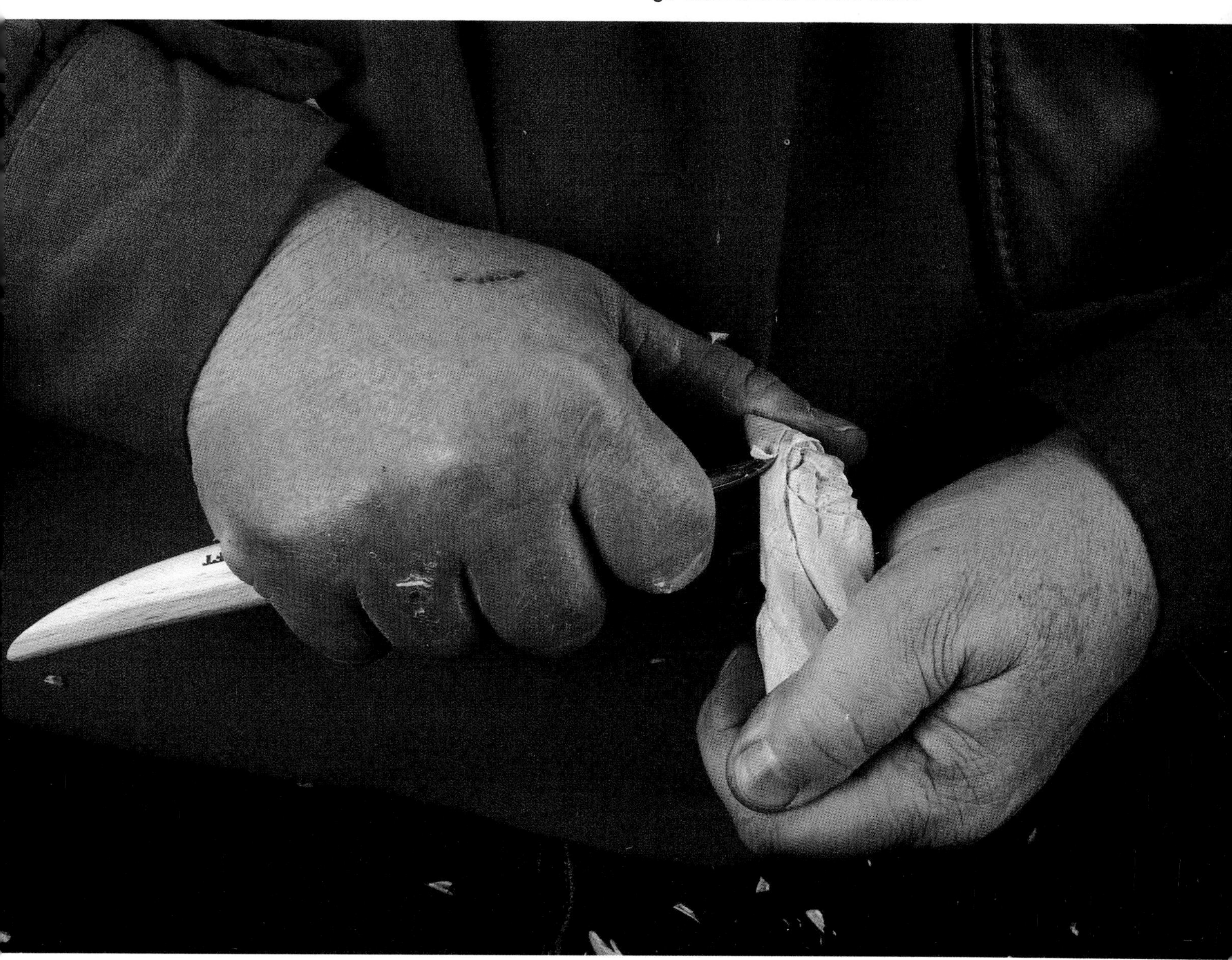

Smooth the carving all over.

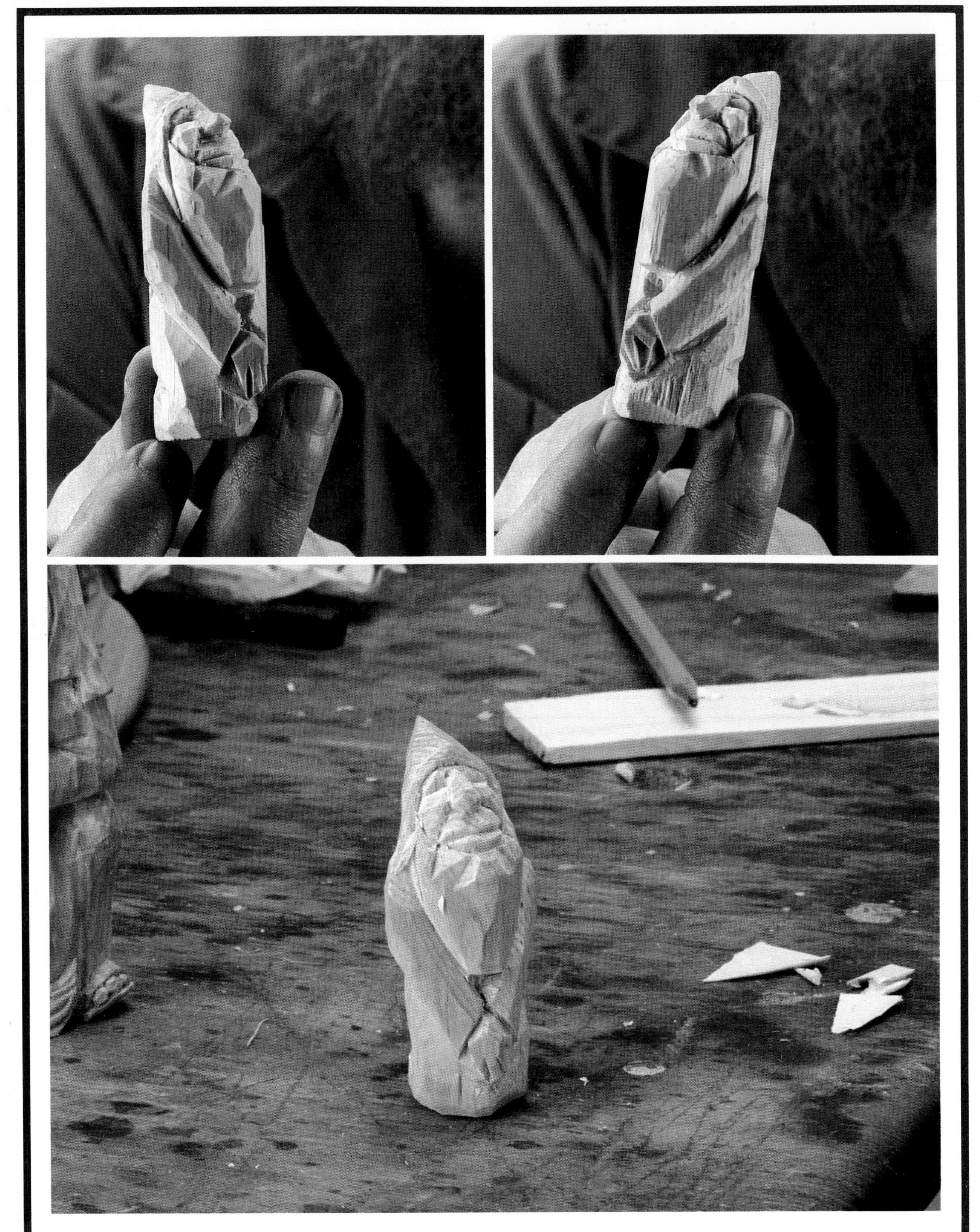

Here is the finished carving.

Painting a Robed Gnome

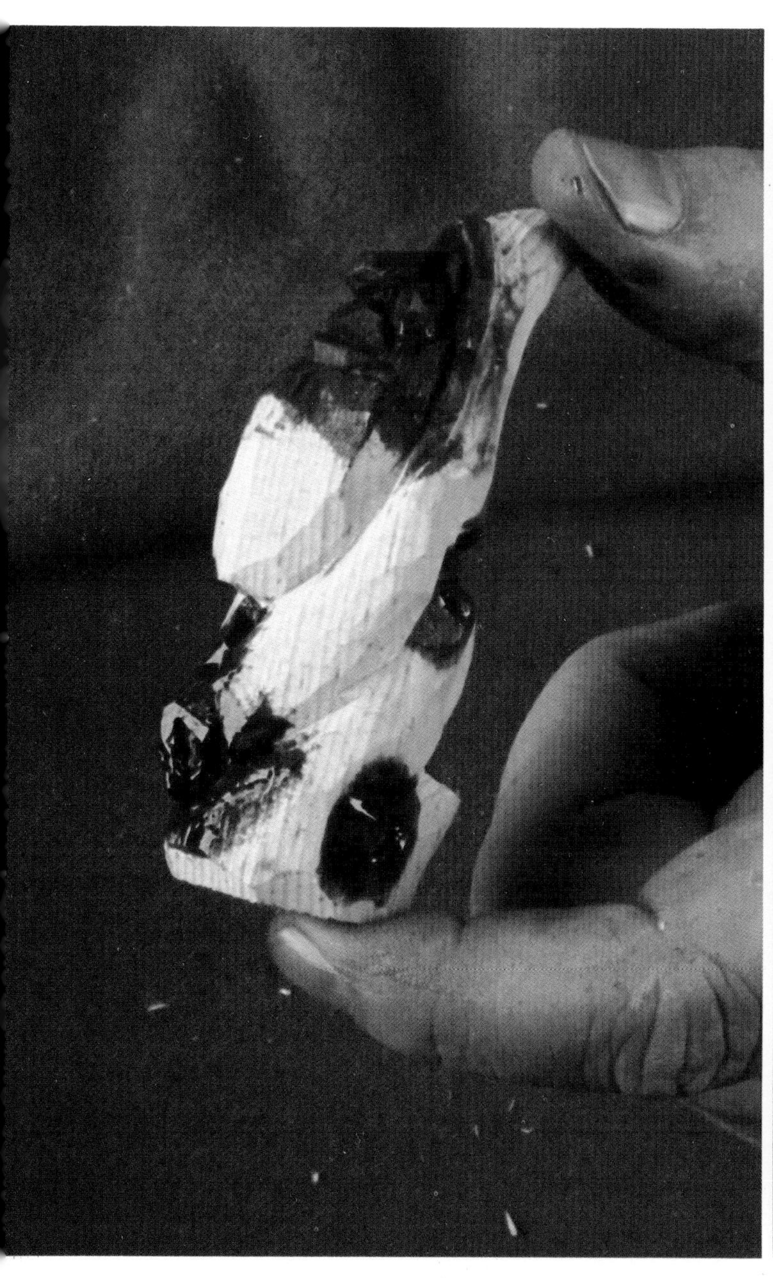

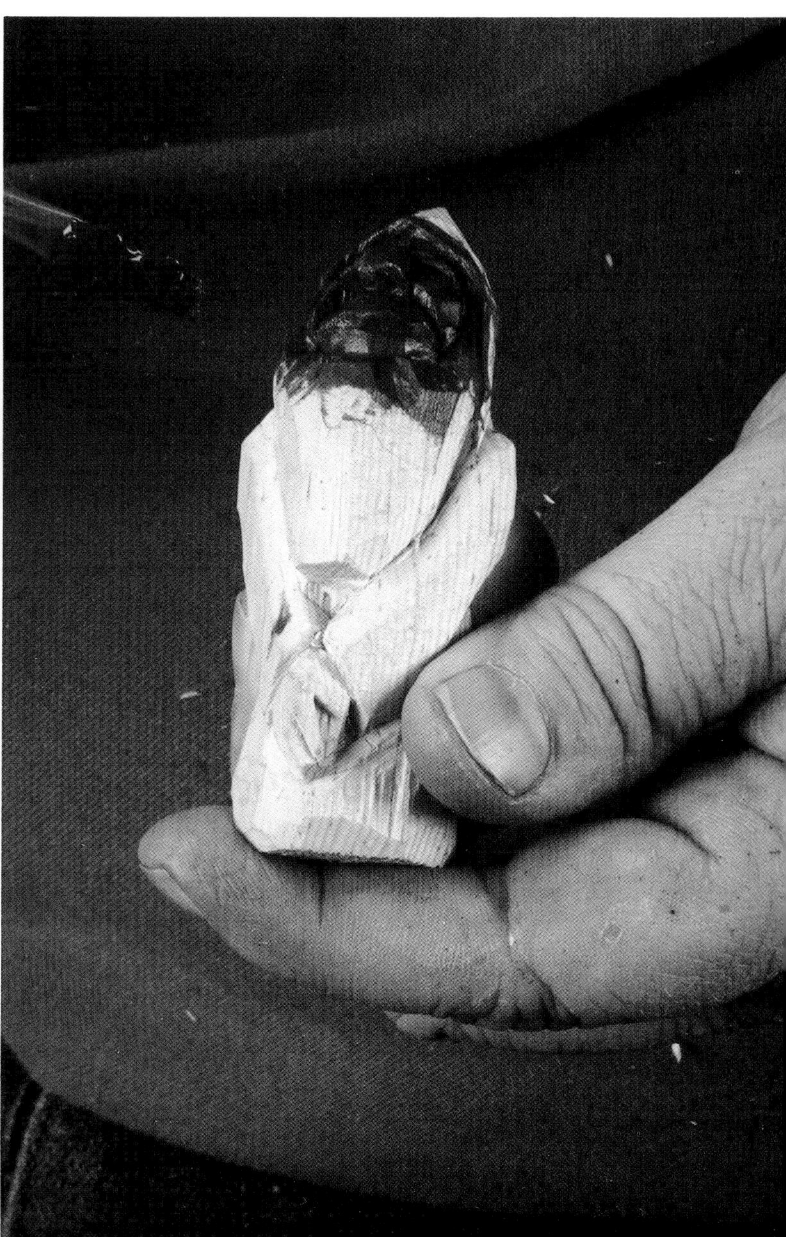

With this carving we will use a diluted burnt sienna as a stain. First apply heavy pigment in the cracks and crevices to bring out the character.

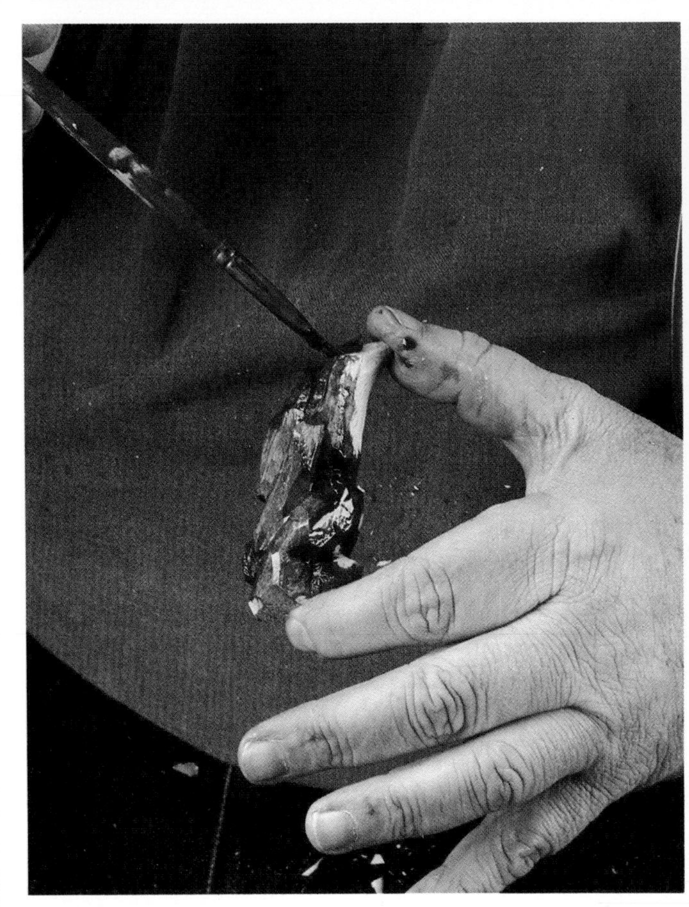

Then use a diluted pigment to wash it out.

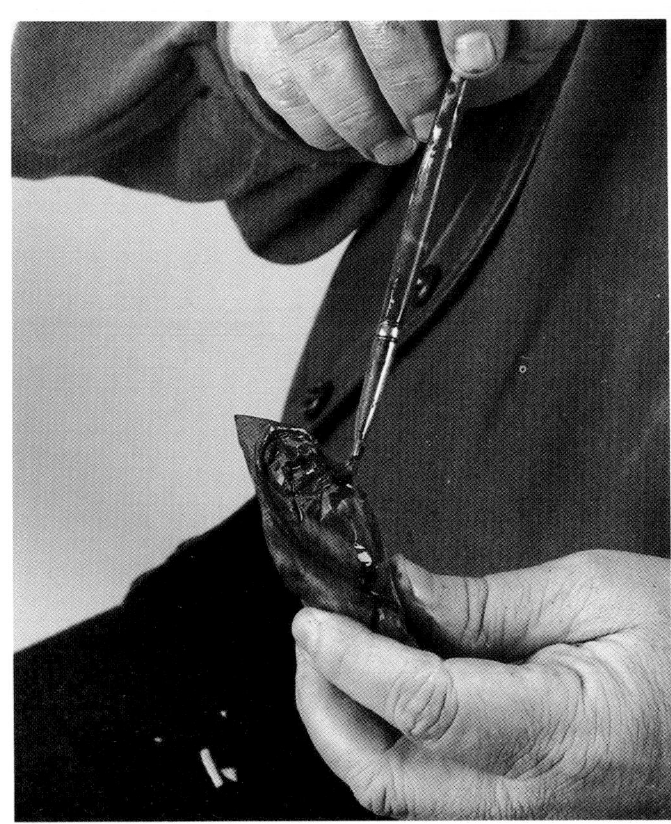

If necessary, use plain turpentine.

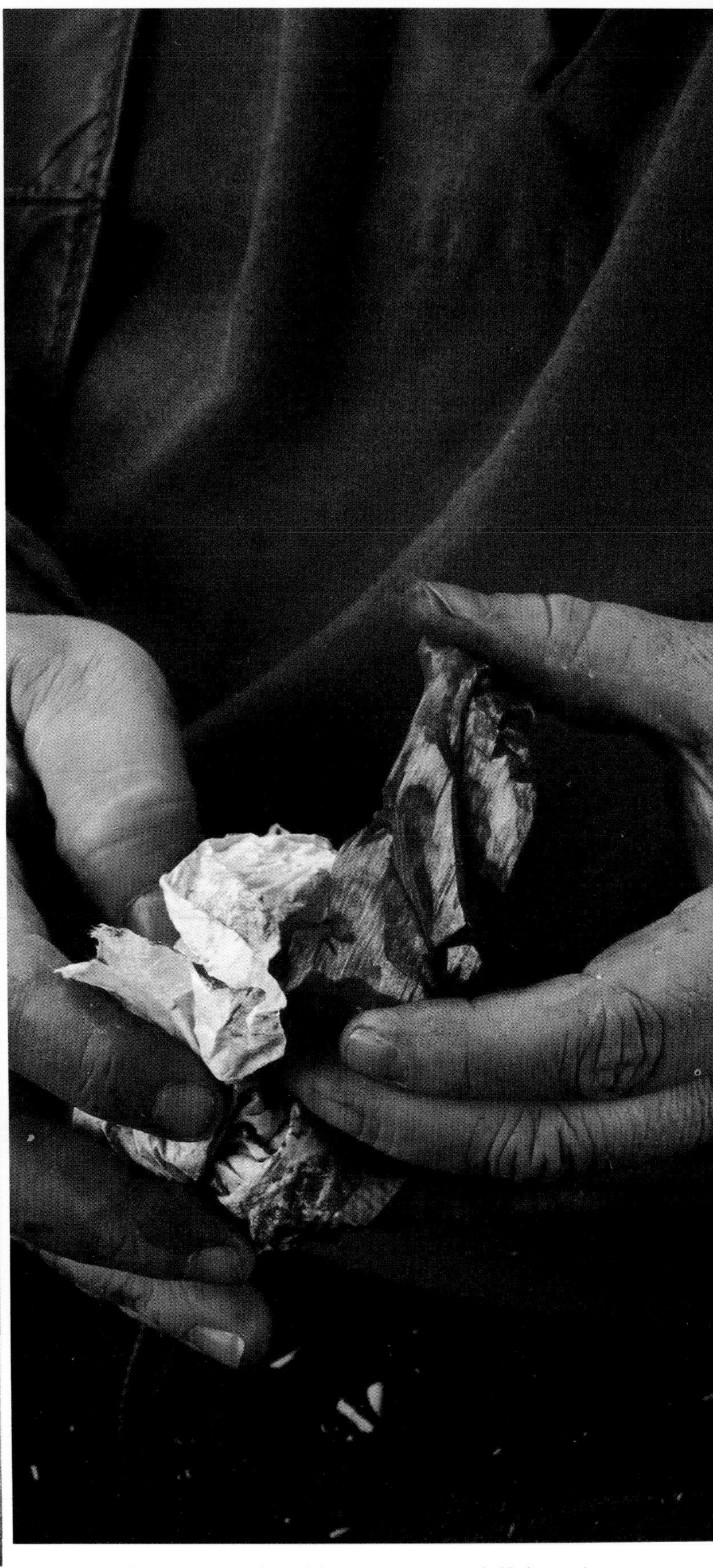

Wipe off excess stain with a paper towel. If there is too much pigment, you may wish to use sandpaper.

The finished product.

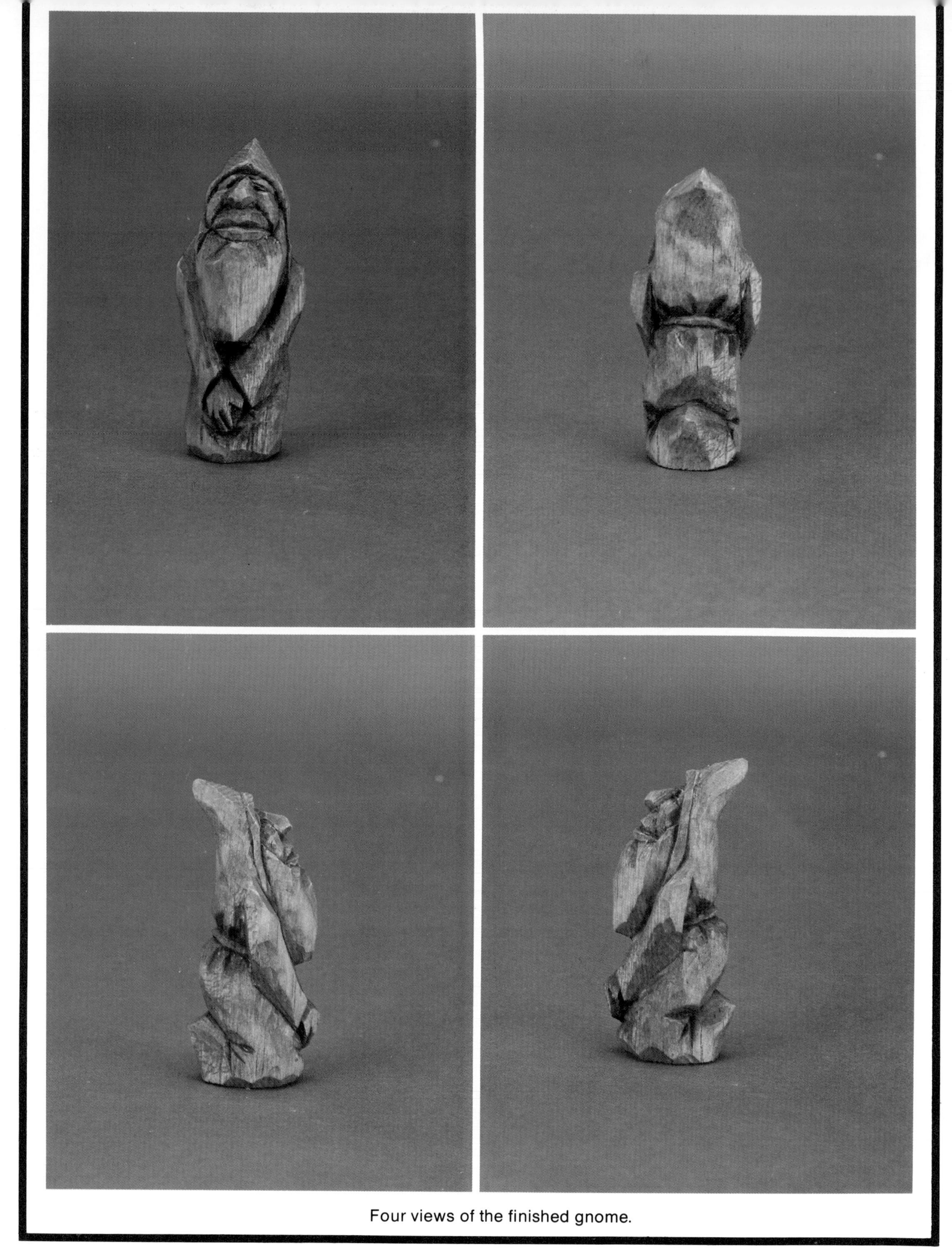

Four views of the finished gnome.

Four views of two unfinished gnome friends

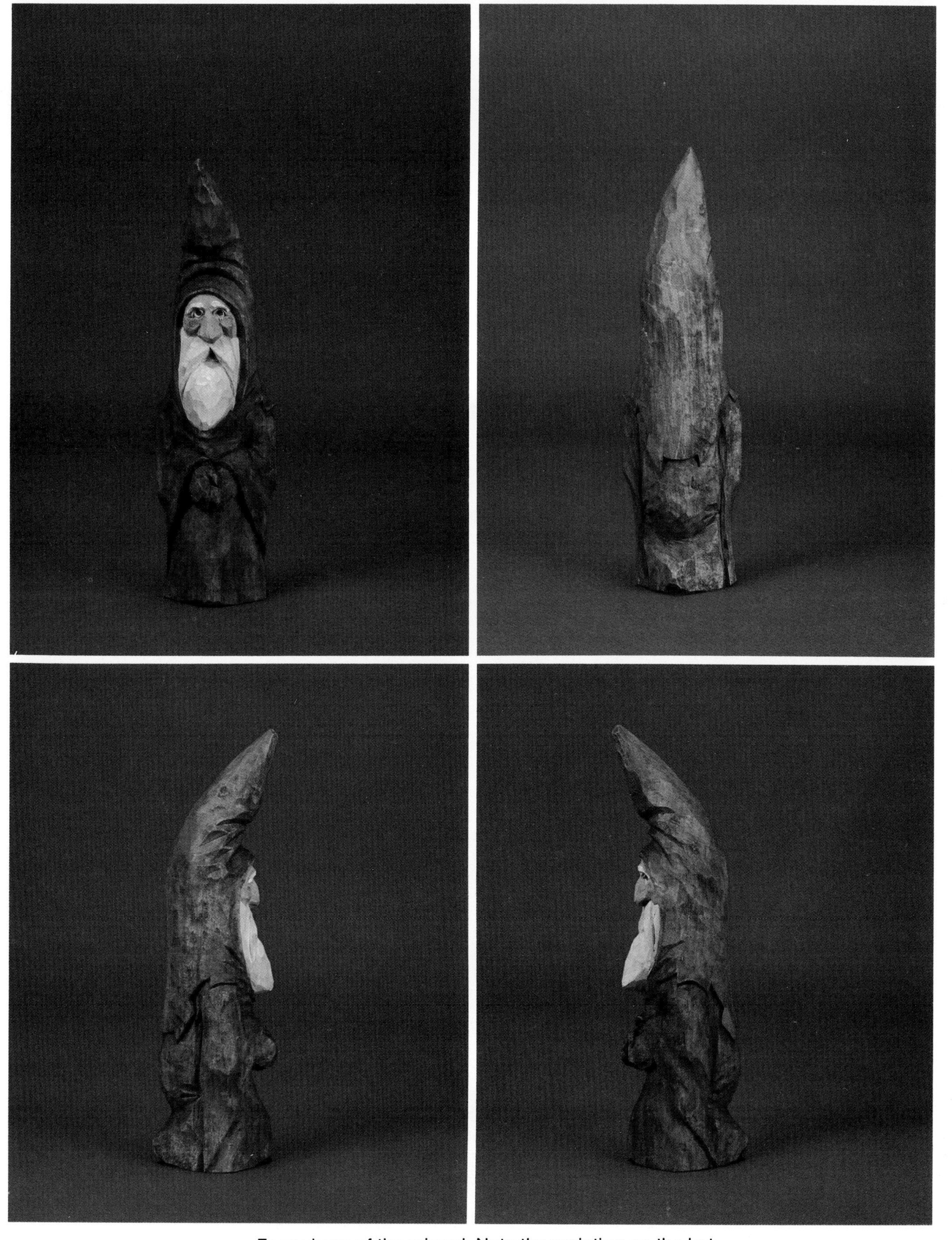

Four views of the wizard. Note the variation on the hat. The corner is the front, not the back, of the hat. This gives it a wholly different character.

Gnomes in a clutch.